Power Electronics Design Handbook

The EDN Series for Design Engineers

N. Kularatna *Power Electronics Design Handbook: Low-Power Components and Applications*
C. Schroeder *Printed Circuit Board Design Using AutoCAD*
EDN Design Ideas (CD-ROM)
J. Lenk *Simplified Design of Voltage-Frequency Converters*
J. Lenk *Simplified Design of Data Converters*
F. Imdad-Haque *Inside PC Card: CardBus and PCMCIA Design*
C. Schroeder *Inside OrCAD*
J. Lenk *Simplified Design of IC Amplifiers*
J. Lenk *Simplified Design of Micropower and Battery Circuits*
J. Williams *The Art and Science of Analog Circuit Design*
J. Lenk *Simplified Design of Switching Power Supplies*
V. Lakshminarayanan *Electronic Circuit Design Ideas*
J. Lenk *Simplified Design of Linear Power Supplies*
M. Brown *Power Supply Cookbook*
B. Travis and I. Hickman *EDN Designer's Companion*
J. Dostal *Operational Amplifiers, Second Edition*
T. Williams *Circuit Designer's Companion*
R. Marston *Electronic Circuits Pocket Book: Passive and Discrete Circuits (Vol. 2)*
N. Dye and H. Granberg *Radio Frequency Transistors: Principles and Practical Applications*
Gates Energy Products *Rechargeable Batteries: Applications Handbook*
T. Williams *EMC for Product Designers*
J. Williams *Analog Circuit Design: Art, Science, and Personalities*
R. Pease *Troubleshooting Analog Circuits*
I. Hickman *Electronic Circuits, Systems and Standards*
R. Marston *Electronic Circuits Pocket Book: Linear ICs (Vol. 1)*
R. Marston *Integrated Circuit and Waveform Generator Handbook*
I. Sinclair *Passive Components: A User's Guide*

Power Electronics Design Handbook

**Low-Power Components
and Applications**

Nihal Kularatna

BOSTON, OXFORD, JOHANNESBURG, MELBOURNE, NEW DELHI, SINGAPORE

Newnes is an imprint of Butterworth–Heinemann.

Copyright © 1998 by Butterworth–Heinemann

 A member of the Reed Elsevier group

All rights reserved.

No part of this publication may be reproduced, stored in a retrieval system, or transmitted in any form or by any means, electronic, mechanical, photocopying, recording, or otherwise, without the prior written permission of the publisher.

Recognizing the importance of preserving what has been written, Butterworth–Heinemann prints its books on acid-free paper whenever possible.

 Butterworth–Heinemann supports the efforts of American Forests and the Global ReLeaf program in its campaign for the betterment of trees, forests, and our environment.

Library of Congress Cataloging-in-Publication Data
Kularatna, Nihal.
    Power electronics design handbook : low-power components and applications / Nihal Kularatna.
       p. cm. — (EDN series for design engineers)
    Includes bibliographical references and index.
    ISBN 0–7506–7073–8 (alk. paper)
    1. Power electronics—Design and construction. 2. Low voltage systems—Design and construction.   I. Title   II. Series
TK7881.15.K85 1998b
621.31' 7—dc21
                                                98–8326
                                                   CIP

British Library Cataloguing-in-Publication Data
A catalogue record for this book is available from the British Library.

The publisher offers special discounts on bulk orders of this book.
For information, please contact:
Manager of Special Sales
Butterworth-Heinemann
225 Wildwood Avenue
Woburn, MA 01801-2041
Tel: 781-960-2500
Fax: 781-960-2620

For information on all Butterworth–Heinemann publications available, contact our World Wide Web home page at: http://www.bh.com

10 9 8 7 6 5 4 3 2 1

Printed in the United States of America

To my family. . . Priyani, Dulsha, and Malsha

Contents

Foreword	xiii
Preface	xv
Acknowledgments	xvii

Chapter 1   Introduction — 1

1.1	Power Electronics Industry	1
1.2	Power Conversion Electronics	2
1.3	Importance of Power Electronics in the Modern World	2
1.4	Semiconductor Components	3
1.5	Power Quality and Modern Components	4
1.6	Systems Approach	6
1.7	Specialized Applications	6
	References	8

Chapter 2   Power Semiconductors — 9

2.1	Introduction	9
2.2	Power Diodes and Thyristors	10
2.3	Gate Turn-Off Thyristors	26
2.4	Bipolar Power Transistors	28
2.5	Power MOSFETs	36
2.6	Insulated Gate Bipolar Transistor (IGBT)	47
2.7	MOS Controlled Thyristor (MCT)	49
	References	52
	Bibliography	53

Chapter 3   DC to DC Converters   55

3.1	Introduction	55
3.2	DC to DC Conversion Fundamentals	56
3.3	Converter Topologies	60
3.4	Control of DC to DC Converters	71
3.5	Resonant Converters	76
3.6	Special DC to DC Converter Designs	84
3.7	DC to DC Converter Applications and ICs	93
3.8	State of the Art and Future Directions	95
References		96
Bibliography		98

Chapter 4   Off-the-Line Switchmode Power Supplies   99

4.1	Introduction	99
4.2	Building Blocks of a Typical High Frequency Off-the-Line Switching Power Supply	100
4.3	Magnetic Components	106
4.4	Output Section	117
4.5	Ancillary, Supervisory and Peripheral Circuits	129
4.6	Power Supplies for Computers	129
4.7	Modular SMPS Units for Various Industrial Systems	130
4.8	Future Trends in SMPS	132
4.9	Field Trouble-Shooting of Computer Systems Power Supplies	133
References		134

Chapter 5   Rechargeable Batteries and Their Management   137

5.1	Introduction	137
5.2	Battery Terminology	137

5.3	Battery Technologies: An Overview	141
5.4	Lead-Acid Batteries	142
5.5	Nickel Cadmium (NiCd) Batteries	148
5.6	Nickel Metal Hydride Batteries	151
5.7	Lithium-Ion (Li-Ion) Batteries	153
5.8	Reusable Alkaline Batteries	156
5.9	Zn-Air Batteries	158
5.10	Battery Management	159
References		173
Bibliography		174

Chapter 6 Protection Systems for Low Voltage, Low Power Systems — 175

6.1	Introduction	175
6.2	Types of Disturbances	176
6.3	Different Kinds of Power Protection Equipment	177
6.4	Power Synthesis Equipment	199
References		199
Bibliography		200

Chapter 7 Uninterruptible Power Supplies — 201

7.1	Introduction	201
7.2	Different Types of Uninterrupted Power Supplies	202
7.3	UPS System Components	211
7.4	UPS Diagnostics, Intelligence, and Communications	223
7.5	UPS Reliability, Technology Changes and the Future	226
References		227
Bibliography		227

Chapter 8 Energy Saving Lamps and Electronic Ballasts — 229

8.1 Introduction — 229
8.2 Gas Discharge Lamps and High Intensity Discharge Lamps — 230
8.3 Introduction to Ballasts — 232
8.4 Some Definitions and Evaluation of Performance — 233
8.5 Conventional Ballasts — 235
8.6 High Frequency Resonant Ballasts — 236
8.7 The Next Generation of Ballasts — 237
8.8 Power Factor Correction and Dimming Ballasts — 239
8.9 Comparison of Compact Fluorescent Lamps Using Magnetic and Electronic Ballasts — 239
8.10 Future Developments of Electronic Ballasts — 241
References — 242

Chapter 9 Power Factor Correction and Harmonic Control — 243

9.1 Introduction — 243
9.2 Definitions — 244
9.3 Harmonics and Power Factor — 245
9.4 Problems Caused by Harmonics — 247
9.5 Harmonic Standards — 248
9.6 Power Factor Correction — 249
9.7 Power Factor Correction ICs — 257
9.8 Active Low Frequency Power Factor Correction — 265
9.9 Evaluation of Power Factor Correction Circuits — 267
References — 268
Bibliography — 268

Chapter 10	**Power Integrated Circuits, Power Hybrids, and Intelligent Power Modules**	**271**
10.1	Introduction	271
10.2	Evolution of Power Integrated Circuits	272
10.3	BCD Technology	274
10.4	Applications of Power Integrated Circuits	275
10.5	Power Hybrids	276
10.6	Smart Power Devices	279
10.7	Smart Power Microcontrollers	284
10.8	System Components and Impact of IGBTs	288
10.9	Future	290

References	**291**

Index	**293**

Foreword

I am pleased to introduce Nihal Kularatna's second book on electronic engineering.

As the Principal Research Engineer of the Arthur C. Clarke Institute for Modern Technologies, Nihal leads a small team which helps adapt modern technologies to suit Sri Lanka's growing needs. The Institute, established by the government in 1983, is devoted to computers, communications, energy, space technologies, and robotics, and serves as the cutting edge for these technologies in Sri Lanka.

Nihal's first book, published two years ago, was well received, and is about to have a new edition. I hope this second book will be a similar success.

Sir Arthur C. Clarke, kt, CBE
Patron, Arthur C. Clarke Institute for Modern Technologies
Chancellor, International Space University
Fellow of King's College, London
Colombo, Sri Lanka

24 April 1998

Preface

During mid-1970s, as a young engineer entering the electronic engineering profession I enjoyed the opportunity to work with processor based online computer systems with no single chip microprocessors (i.e. processor systems designed with the basic TTL family), and the early generations of navigational aids based on basic analog and digital components. This work gave me the opportunity to play with bulky linear power supplies and UPS systems etc. which made me appreciate the problems with the commercial power supply interface. In the early 1980s, I spent several years with digital telephone exchanges which had both microprocessors and high speed logic based PCBs. Some of these experiences showed me that the power electronic interface, which converts the AC power supply to low voltage DC systems, needs very special attention during design, maintenance and management, particularly catering for transients such as spikes and surges.

With the power semiconductors and integrated circuits maturing over the two and half decades from 1972 (when the first microprocessor was released), middle-aged engineers like us were able to observe and appreciate the development of new power electronics techniques for switchmode power supplies, UPS systems, surge suppressers, battery management circuitry and energy saving lighting etc. By the mid-1990s reasonably matured design techniques were available for tackling the power supply interface of the complex systems designed with modern VLSI and ULSI components of sub-micron feature sizes. During the last decade power electronics industry sector was growing rapidly. As highlighted in an editorial in PCIM Journal, industry related to power electronics is rapidly growing and it is almost accounting for over 1/20 of the total electronics industry in the United States.

As a result of the unprecedented demand for power electronic subsystems, many modern components and design techniques were developed by the dedicated researchers, industry engineers and other professionals during the last two decades. Power semiconductors—the muscle of the power electronic products have gained unprecedented voltage and current handling capabilities, while the controller chips—the heart and the brain of the power electronic subsystems are becoming more intelligent or smart.

Although several theoretical presentations on power electronics topics are available in circulation, there are hardly any practical handbooks that cover some of the latest components, techniques and applications. Getting involved with the design and development of power conditioning and inverter techniques, as well as conducting short courses on power electronics helped me to develop a practical information base for the manuscript of a book which could fill this gap. My task was not to cover the broad world of power electronics, but to limit the contents to what is encountered in modern information processing environments as well as in the energy saving lighting, power factor correction blocks, etc. as selective topics.

I have attempted my best to present the state of the art on most topics by discussing the commercially available components, design trends and the applications in these selected topics, without attempting to cover all areas in power electronics. My readers are invited to judge how I have performed in this task and your comments are most welcome.

A. D. V. Nihal Kularatna
39A Sumudu Place
Sri Rahula Road
Katubedda
Moratuwa
Sri Lanka

24 March 1998

Acknowledgments

In carrying out this exercise while living on the island of Sri Lanka where the electronic industry is still developing, and only in very limited areas, I know that the task I have attempted would have been impossible without the support and encouragement of my many friends, colleagues and home/office support staff. Many manufacturers of semiconductors and power electronic products in the U.S. and Europe kindly provided me necessary industrial and design information for relevant chapters.

I am grateful to the following companies, and their technical experts, who provided information and permission for material used in respective chapters:

- Intel Corporation and PCIM editorial staff for information in Chapter 1.
- Dr. Colin Rout of GEC Plessy Semiconductors, UK; Edward Pawlak of Harris Semiconductors, USA; Brian Goodburn of Motorola, USA; and Michael T. Robinson of International Rectifier for information in Chapter 2.
- Maxim Integrated Products Inc., and Linear Technology Corporation for information from their data books, articles and application notes for the benefit of Chapter 3.
- Magnetics Inc., Power Trends Inc., and Unitrode Integrated Circuits for information in Chapter 4.
- Ms. Patty Smith of Benchmarq Microelectronics, USA, Tim Cutler of AER Energy Resources, and Pauline Tonkins of Moli Energy for information and permission to use material for items in Chapter 5.
- Richard Zajkowski of Liebert Corporation, USA, for most useful information and suggestions related to Line Interactive UPS systems and Sam Wheeler of Power Quality Assurance Journal advisory board for information in Chapter 7.
- Peter N. Wood and Ajit Dubhashi of International Rectifier, USA, and Micro Linear Corporation, USA, for information utilized in Chapter 8.
- Unipower Corporation, Motorola, Micro Linear Corporation for information in Chapter 9.
- Dr. Rich Fassler and Fritz Lns of IXYS Corporation, Ms. Karen Banks of Motorola Inc., Ms. Marie G. Rivera of Apex Microtechnology, Powerex Inc., SGS–Thomson Semiconductors, Cherry Semiconductor Corporation, and Solitron Inc., for information found in Chapter 10. Also, my most kind gratitude is extended to Larry Spaziani (formerly of Unitrode) and Ms. Claire Taylor (formerly of Harris Semiconductor) and Frank Wahl of PCIM Journal for their assistance and support.

Without support from the worldwide power electronics industry this book might have, wrongly been weighted more towards the academic rather than the practical for which it was intended. Additionally, I am grateful to my friend Wayne Houser, retired Voice of America Foreign Service officer, who provided valuable liaison assistance and moral support during the information collection process from his office in California.

In preparation of the manuscript text I am grateful to secretaries Chandrika Weerasekera, Indrani Hewage, Dilkusha de Silva and Neyomi Fernando.

For the creation of computer graphics in figures and graphs in the chapters I am grateful to Thilina Wijesekera and Arosh Edirisinghe. Additional assistance was provided by Kapila Kumara, Promod Hettihewa and Chandana Amith. Without their loyal and dedicated assistance the project would have become an impossible task.

Many students enrolled in my Continuing Professional Development (CPD) courses at the Institute helped me develop the basis for this manuscript. With pleasure I also acknowledge the great service provided by the editorial staff of industry journals including *EDN, PCIM, PQ Assurance, Electronic Design,* and *EPE Journal*, which were my sources for reliable and current information. I am also very grateful to Mr. Padmsiri Soysa, and his staff, at the ACCIMT library who loyally research, collect and continually make available to the staff and the public, the Institute's technical information resources.

I am also grateful to my friends in local industry, such as Keerthi Kumarasena and others for their continuous assistance. For computing resources, and the maintenance of same in a very timely manner, I am thankful to the Managing Directors of Metropolitan Group, JJ Amabani and DJ Ambani, Niranjan de Silva, Director, Mohan Weerasuriya, Senior Service Manager, and their staff.

Also, I appreciate the encouragement given to my work by Mr. S. Rubasingam, Librarian, University of Moratuwa, Sherani Godamunne, Shantha and Jayantha De Silva and my relatives and many friends. Feedback I received from Dr. Robin Mellors-Bourne and his reviewers at IEE publishing helped me improve the manuscript and I am very thankful to them.

I can never forget the assistance provided by my friend, Dr. Mohan Kumaraswamy in 1977 at the time I wrote my first design article, which was the trigger for my subsequent publishing efforts. Four chapter co-authors, Drs. Dileeka Dias, Aruna Ranweera, Nalin Wickramarchchi and Mr. Anil Gunawardana helped create a team spirit in this work and I thank each of them for their contributions to this book. I am very thankful to Ms. Josephine Gilmore, and the staff of Butterworth–Heinemann, with special gratitude to Pam Chester and Susan Prusak for their speedy schedule related to the publication.

Former chairman of the Arthur C. Clarke Institute for Modern Technologies-Professor K.K.Y.W. Perera, Professor Sam Karunaratne, Director and Chairman of the Board of Governors of ACCIMT provided specific encouragement for this work which is very much appreciated.

Warmly I thank the institution's patron, Sir Arthur C. Clarke, and Dr. Frederick C. Durant III, Executive Director of the Arthur C. Clarke Foundation USA, for their continuing encouragement of my work. Sir Arthur also kindly provided the forward for this book for which I am additionally grateful.

Nihal Kularatna
Computer Division
Arthur C Clarke Institute for Modern Technologies
Katubedda, Moratuwa
Sri Lanka

26 April 1998

CHAPTER **1**

Introduction

1.1 Power Electronics Industry

Utility systems usually generate, transmit, and distribute power at a fixed frequency such as 50 or 60 Hz, while maintaining a reasonably constant voltage at the consumer's terminal. The consumer may use many different electronic or electrical products which consume energy from a DC or AC power supply which converts the incoming AC into the required form.

In the case of products or systems running on AC, the frequency may be the same, higher, lower or variable compared to the incoming frequency. Often, power needs to be controlled with precision. A power electronics system interfaces between the utility system and consumer's load to satisfy this need.

The core of most power electronic apparatus consists of a converter using power semiconductor switching devices that works under the guidance of control electronics. The converters can be classified as rectifier (AC-to-DC converter), inverter (DC-to-AC converter), DC-to-DC converter, or an AC power controller (running at the same frequency), etc.

Often, a conversion system is a hybrid type that mixes more than one basic conversion process. The motivation for using switching devices in a converter is to increase conversion efficiency to a high value. In few situations of power electronic systems, the devices (power semiconductors) are used in the linear mode too, even though due to reasons of efficiency it is getting more and more limited.

Power electronics can be described as an area where anything from a few watts to over several hundred megawatt order powers are controlled by semiconductor control elements which consume only few microwatts to milliwatts in most areas. As per industry estimates indicated in an editorial of *Power Conversion and*

Intelligent Motion Journal (1995), the power electronics industry component in the U.S. was around US$ 30 billion, from a total estimated electronics industry of around US$ 570 billion.

1.2 Power Conversion Electronics

Power conversion electronics can be described as a group of electrical and electronic components arranged to form an electric circuit or group of circuits for the purpose of modifying or controlling electric power from one form to another. For example, power conversion electronics is employed to provide extremely high voltages to picture tubes to display the courses of aircraft approaching an airport.

In another example, power conversion electronics is employed to step up low voltage from a battery to the high voltage required by a vacuum fluorescent display to allow paramedics to display a victim's heartbeat on a screen. This also allows paramedics to gain information en route to the hospital, which may save the patient's life.

Twenty years ago, power conversion was in its infancy. High efficiency switchmode power supplies were a laboratory curiosity, not a production line reality. Complex control functions, such as the precision control of stepper motors for robotics, microelectronics for implanted pacemakers, and harmonic-free switchmode power supplies, were not economically achievable with the limited capabilities of semiconductor components available at the time.

1.3 Importance of Power Electronics in the Modern World

At the beginning of the 20th century the world population was around 1.5 billion; by the year 2000 it is projected to be around six billion. Rapid technology evolution coupled with the population explosion has resulted in an increase in average electrical power usage, from about one-half million MW in the year 1940 to a projected five million MW in the year 2000. This magnitude of growth—when coupled with the increasing electrical power sophistication associated with process control, communications, consumer appliances/electronics, information management, electrified transportation, medical, and other applications—results in roughly 45 percent of all electrical power delivered to user sites today being reprocessed via power electronics. This is expected to increase to about 75 percent by the year 2000. By the turn of the century approximately 3.8 million MW of electrical power will be processed by power electronics (Marcel, Gaudreav, Wieseneel, and Dionne 1993).

Typical power electronics applications include electronic ballasts, high voltage DC transmission systems, power conditioners, UPS systems, power supplies, motor drives, power factor correction, rectifiers and, more recently, electric vehicles. With computer systems, telecom products and a plethora of electronic consumer appliances which require many power electronic sub-systems, the power electronics industry has become an important topic in the electronics industry and the information technology area.

1.4 Semiconductor Components

In modern power electronics apparatus, there are essentially two types of semiconductor elements: the power semiconductors that can be defined as the muscle of the equipment, and microelectronic control chips, which provide the control and intelligence. In most situations operation of both are digital in nature. One manipulates large power up to mega or gigawatts, the other handles power only on the order of microwatts to milliwatts.

Until the 1970s, power semiconductor technology was based exclusively upon bipolar devices, which were first introduced commercially in the 1950s. The most important devices in this category were the p-i-n power rectifier, the bipolar power transistor, and the conventional power thyristor. The growth in the ratings of these devices was limited by the availability of high purity silicon wafers with large wafer diameter, and their maximum switching frequency was limited by minority carrier lifetime control techniques. In the 1980s another bipolar power device, the Gate Turn-Off Thyristor (GTO), became commercially available with ratings suitable for very high power applications. Its ability to turn-on and turn-off large current levels under gate control eliminated the commutation circuits required for conventional thyristors, thus reducing size and weight in traction applications, etc.

Although these bipolar power devices have been extensively used for power electronic applications, a fundamental drawback that has limited their performance is the current controlled output characteristic of the devices. This characteristic has necessitated the implementation of high power systems with powerful discrete control circuits, which are large in size and weight.

In the 1970s, the first power Metal-Oxide-Semiconductor Field Effect Transistors (MOSFETS) became commercially available (Severns and Armijos 1984). Their evolution represents the convergence of power semiconductor technology with mainstream CMOS integrated circuit technology for the first time. Subsequently, in the 1980s, the Insulated Gate Bipolar Transistor (IGBT) became commercially available.

The MOSFET and IGBT require negligible steady state control power due to their Metal-Oxide-Semiconductor (MOS) gate structure. This feature has made them extremely convenient for power electronic applications resulting in a rapid growth in the percentage of their market share for power transistors.

The ratings of the power MOSFET and IGBT have improved rapidly in recent years, resulting in their overtaking the capability of bipolar power transistors. The replacement of bipolar power transistors in power systems by these devices that was predicted a decade ago has now been confirmed. However, the physics of operation of these devices limits their ability to handle high current levels at operating voltages in excess of 2000 volts.

Consequently, for high power systems, such as traction (electric locomotives and trams) and power distribution, bipolar power devices, namely the thyristor and GTO, are the best commercially available components today. Although the power ratings for these devices continue grow, the large control currents needed to switch the GTOs has stimulated significant research around the world aimed at the development of MOS-gated power thyristor structures such as MOS Controlled Thyristors (MCT).

The development of the insulated gate power devices discussed above has reduced the power required for controlling the output transistors in systems. The relatively small (less than an ampere) currents at gate drive voltages of less than 15 volts that are needed for these devices can be supplied by transistors that can be integrated with CMOS digital and bipolar analog circuitry on a monolithic silicon chip. This led to the advent of smart power technology in the 1990s.

Smart power technology provides not only the control function in systems but also serves to provide over-current, over-voltage, and over-temperature protection, etc. At lower power levels, it enables the implementation of an entire sub-system on a monolithic chip. The computer-aided design tools that are under development will play an important role in the commercialization of smart power technology because they will determine the time-to-market as well as the cost for development of Power Application Specific Integrated Circuits (PASIC). Sometimes these devices are called Application Specific Power Integrated Circuits (ASPIC).

In systems such as automotive electronics or multiplex bus networks and power supplies for computers with low operating voltages (below 100 volts), the power MOSFET provides the best performance. In systems such as electric trams and locomotives, the GTO is the best commercially available component. In the near future, MOS gated thyristor structures are likely to replace the GTO.

Towards the mid-90s GaAs power diodes have entered the marketplace providing better switching characteristics as well as lower forward drop, etc. On a longer time frame, it is possible that devices based upon wide band-gap semiconductors, such as Silicon Carbide, could replace some of these silicon devices.

1.5 Power Quality and Modern Components

During the last decade, many industrial processes have dramatically expanded their use of electronic equipment with very sophisticated microelectronic components. Meanwhile many consumer electronic products and personal computers, etc. are used by many millions of individuals at their residences too. Downsizing of individual semiconductor components in processor and memory chips, is evident from the exponential growth in the number of components per chip in popular microprocessors such as the Intel family (Figure 1–1).

With the downsizing of semiconductors in the components, the quality of AC power systems becomes critical for the reliable operation of the products and systems. Common problems such as blackouts, brownouts, sags, spikes, and lightning related transients, etc. propagating into the systems could create serious problems in the systems. With more and more nonlinear subsystems (such as switchmode power supplies, switched rectifiers, etc.) used at the interface between utility power input and the systems, the power quality problem is worsened due to the non-linear nature of the currents drawn from the utility.

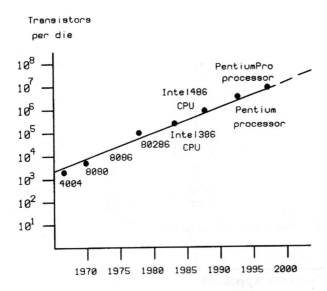

FIGURE 1-1 Exponential growth in number of transistors per chip in Intel processor family (Reproduced by permission of Intel Corp, USA)

For this reason, power factor correction, and harmonic control, etc., which were specifically relevant to the electrical power environments historically, are now becoming mandatory with low power systems too. Many organizations, such as component manufacturers, system designers as well as standardization groups are placing heavy emphasis on these concepts.

Power factor corrected switchmode power supplies, power factor corrected energy saving lamps, etc. are gradually becoming the modern design trends. AC voltage regulators, power conditioners and UPS systems, etc. are becoming a very fast growing market segment due to power quality issues.

While these systems use the state of the art power semiconductors, etc., highly sophisticated systems with superconducting magnetic energy storage (SMES), etc. are also installed as trial systems in critical locations (De Winkel, Loslenben, and Billmann 1993). Superconducting magnetic energy storage was originally proposed for use by utilities to store energy to meet peak electricity demands.

The systems were to store large amounts of electrical energy to provide thousands of megawatts of power for several hours at a time. However, smaller storage systems have developed much faster. The first commercially available unit of this kind rapidly stores and delivers smaller amounts of electricity over a brief period (about a megawatt for a few seconds). This new technology excels in handling power disturbances, which are increasingly expensive problems in industrial facilities.

6 POWER ELECTRONICS DESIGN HANDBOOK

FIGURE 1-2 Basic power electronic system

1.6 Systems Approach

Taking a systems approach, every application in power electronics can be represented by a load, and a drive consisting of four basic elements (Figure 1–2), control - driver - device - circuit. Most of the progress in the near future is expected to originate either from the side of the control unit (i.e., more sophisticated software) or from the device side. In both cases, advances in silicon technology open up new possibilities, especially in order to cut costs for a given application.

Another area where a large effort is being spent, both in research and development, concerns the combination of these building blocks. In the first stage, this includes the integration of the driver into the actual device. This effort is presently under way in the sense of replacing the conventional current fed drive by a voltage fed one.

Ultimately, the integration of all three blocks, control - driver - device, is the goal. This applies mainly to the region of low power where dissipation problems are less severe.

In the case of industrial applications of power electronics, the operating environment will impose additional requirements beyond the mere functional performance. Qualities like reliability, safety, maintainability, and availability need to be considered too and will influence both the design and the selection of components for a given system.

1.7 Specialized Applications

Frequently, new applications for power electronics are being suggested and created. Some are extensions of other fields; others are fields unto themselves. Here are several new applications that are becoming a reality:

- Magnetically levitated (MAGLEV) trains with advanced electromagnetic propulsion and power systems.
- Plasma fusion technology with very high power electronics systems.
- Megawatt amplifiers in small sizes.
- Smart power management and distribution systems for high speed fault detection and power re-routing.

The capability of superconductors to generate a large magnetic field allows a MAGLEV train to levitate a passenger vehicle above a track so that physical contact is not needed. Slowed only by air friction and track coil resistance, the train can then travel at speeds approaching 500 Km/h. Mechanical propulsion is difficult without physical contact or noisy turbines. Therefore, the train must be propelled electromagnetically. Similarly, power to the passenger compartment must be inductively coupled from the guideway.

Controlled thermonuclear plasma fusion systems require regulated high power supplies, to heat hydrogen isotopes to temperatures that will initiate fusion reactions. Examples of these heating systems are gyrotrons and neutral beams. Both systems require modulation of their input supplies. The power requirements of these systems are staggering; hundred of kilovolts at hundreds of amperes for several seconds at a time. Novel solid state power electronics provides a way to attain higher power with lower cost, easier maintenance, greater ruggedness, and smaller size. A series modulator using IGBT-based switching modules can provide discrete modulation with sub-microsecond switching times. They also are not subject to parasitic oscillations and x-rays, so higher voltages up to the megavolt level are possible.

Some existing industrial applications require amplifiers that can provide megawatt power levels. For example, amplifiers that produce ± 450V at 2,000 A to drive a large electromagnetic linear motor have been reported (Marcel, Gaudreav, Wieseneel, and Dionne 1993). It consists of six parallel, 400A, 1200V IGBTs in each leg of an H-bridge and operates at up to 10KHz switching frequency. Active overcurrent and overvoltage regulation and protection are provided with the amplifier. These power amplifiers can be used for applications such as seismic exploration. References 2 and 3 provide some details related to very special applications.

Advanced power management and distribution is undergoing change as solid state power controllers become available. These devices can be circuit breakers, relays, and power controllers all in one package. They can operate three to four orders of magnitude faster than mechanical circuit breakers and relays. The speed and easy interface with other electronics of these devices offer capabilities far beyond those of mechanical circuit breakers and relays. Fast turn-off gives solid state power controllers the ability to limit current by relying on the inductance inherent in any power distribution system. Easy interface with other electronics, such as microprocessors, makes smart operation and remote control possible. A microprocessor can use information from a variety of sensors to decide the status of the power controller.

This chapter has provided an overview about the power electronics industry, including high power and specialized applications. The next chapters provide the details related to power electronic components and techniques used in power electronic systems handling power in the range of few watts to about 100 kw.

References

1. Thollot, Pierre A. "Power Electronics Technology & Applications." *IEEE*, 1993.
2. De Winkel, Carrel C., James P. Losleben, and Jennifer Billmann. "Recent Applications of Super Conductivity Magnet Energy Storage." *Power Quality Proceedings (USA)*, October 1993, pp 462–469.
3. Marcel, P. J., P. E. Gaudreav, Robert A. Wiesenseel, and Jean-Paul Dionne. "Frontiers in Power Electronics Applications." *Power Quality Proceedings*, October 1993, pp 796–801.
4. Severns, Rudy and Jack Armijos. "MOSPOWER Applications Handbook." Siliconix Inc., 1984.

CHAPTER **2**

Power Semiconductors

2.1 Introduction

During the elapsed half century from the invention of the transistor, the power electronics world has been able to enjoy the benefit of many different types of power semiconductor devices. These devices are able to handle voltages from few volts to several kilovolts, switching currents from few milliamperes to kiloamperes. Within a decade from the invention of transistor, the thyristor was commercialized.

Around 1968 power transistors began replacing the thyristors in switchmode power systems. The power MOSFET, as a practical commercial device, has been available since 1976. When "smart power" devices appeared in the market, designers were able to make use of the Insulated Gate Bipolar Transistor (IGBT) from the early 90s. In 1992 MOS controlled thyristors were commercially introduced and around 1995, semiconductor materials such as GaAs and Silicon Carbide (SiC) have opened new vistas for better performance power diodes for high frequency switching systems.

Presently, the spectrum of what are referred to as "power devices" spans a very wide range of devices and technology. Discrete power semiconductors will continue to be the leading edge for power electronics in the 1990s. Improvements on the fabrication processes for basic components such as diodes, thyristors, and bipolar power transistors have paved way to high voltage, high current, and high speed devices. Some major players in the industry have invested in manufacturing capabilities to transfer the best and newest power semiconductor technologies from research areas to production.

Commercially available power semiconductor devices could be categorized in to several basic groups such as, diodes, thyristors, bipolar junction power transistors (BJT), power metal oxide silicon field effect transistors (Power MOSFET), insulated gate bipolar transistors (IGBT), MOS controlled thyristors (MCT), and gate turn off thyristors (GTO), etc. This chapter provides an overview of the characteristics, performance factors, and limitations of these device families.

2.2 Power Diodes and Thyristors

2.2.1 Power Diodes

The diode is the simplest semiconductor device, comprising a P-N junction. In attempts to improve its static and dynamic properties, numerous diode types have evolved. In power applications diodes are used principally to rectify, that is, to convert alternating current to direct current. However, a diode is used also to allow current freewheeling. That is, if the supply to an inductive load is interrupted, a diode across the load provides a path for the inductive current and prevents high voltages $L\frac{di}{dt}$ damaging sensitive components of the circuit.

The basic parameters characterizing the diodes are its maximum forward average current $I_{F(Ave)}$ and the peak inverse voltage (PIV)1. This parameter is sometimes termed as blocking voltage (V_{rrm}). There are two main categories of diodes, namely "general purpose P-N junction rectifiers" and "fast recovery P-N junction rectifiers." General purpose types are used in circuits operating at the line frequencies such as 50 or 60 Hz. Fast recovery (or fast turn-off) types are used in conjunction with other power electronics systems with fast switching circuits.

Classic examples of the second type are switchmode power supplies (SMPS) or Inverters, etc. Figure 2–1(a) indicates the capabilities of a power device manufacturer catering for very high power systems and Figure 2–1(b) indicates the capabilities of a manufacturer catering for a wide range of applications.

At high frequency situations such as Inverters and SMPS, etc., two other important phenomena dominate the selection of rectifiers. Those are the "forward recovery" and the "reverse recovery."

2.2.1.1 Forward Recovery

The turn-on transient can be explained with Figure 2–2. When the load time constant L/R is long compared to the time for turn on t_{fr} (**forward recovery** time) load current will hardly change during this period. For the time t< 0, the switch Sw is closed. Steady conditions prevail and the diode D is reverse biased at $-V_s$. It is in the off-state, and $i_D = 0$.

At t=0, the switch Sw is opened. The diode becomes forward biased, provides a path for the load current in R and L, so that the diode current i_D rises to I_F ($\approx I_l$) after a short time t_r (**rise time**) and the diode voltage drop falls to its steady value after a further time t_f (**fall time**). This is shown in Figure 2–2(b). The diode turn-on time is the time t_{fr}, that comprises $t_r + t_f$. It takes this time t_{fr} for charge to change from one equilibrium state (off) to the other (on).

Power Semiconductors 11

(a)
(b)

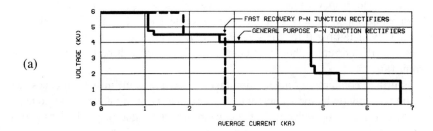

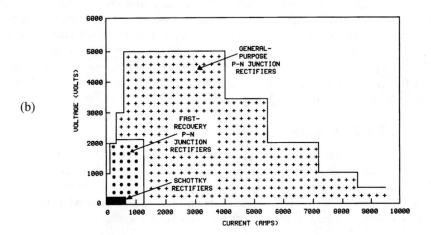

FIGURE 2-1 Rectifier capabilities (a) Rectifier capabilities of a major supplier of high power semiconductors (Reproduced by permission of GEC Plessey Semiconductors, UK.) (b) Rectifier capabilities of a manufacturer catering for wide range of applications (Reproduced by permission of International Rectifier Inc., USA.)

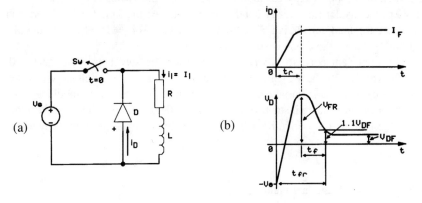

FIGURE 2-2 Turn on characteristics (a) Circuit (b) Waveforms

The total drop V_D reaches a peak forward value V_{FR} that may be from 5 to 20V, a value much greater than the steady value V_{DF} generally between 0.6 to around 1.2V. The time t_r for the voltage to reach V_{FR} is usually about 0.1µs. At a time $t > t_r$, the current i_D becomes constant at I_1 (which will be the forward diode current I_F).

Further, conductivity modulation takes place due to the growth of excess carriers in the semiconductor accompanied by a reduction of resistance. Consequently, the $i_D R_D$ voltage drop reduces. In the equilibrium state, that may take a time of t_f, with a uniform distribution of excess carriers, the voltage drop v_D reaches its minimum steady-state value V_{DF}.

During the turn-on interval t_f, the current is not uniformly distributed so the current density can be high enough in some parts to cause hot spots and possible failure. Accordingly, the rate of rise of current di_D/dt should be limited until the conduction spreads uniformly and the current density decreases. Associated with the high voltage V_{FR} at turn-on, there is high current, so there is extra power dissipation that is not evident from the steady-state model. The turn-on time varies from a few ns to about 1ms depending on the device type.

2.2.1.2 Reverse Recovery

Turn-off phenomenon can be explained using Figure 2–3. In Figure 2–3(a) except for the diode, the circuit elements of this simple chopper are considered to be ideal. Switching Sw at a regular frequency, the source of constant voltage, Vs maintains a constant current I_1 in the RL load, because it is assumed that the load time constant $\frac{L}{R}$ is long compared with the period of the switching.

While the switch Sw is closed, the load is being charged and the diode should be reverse biased. While the switch Sw is open, the diode D provides a freewheeling path for the load current I_1. The inductance L_s is included for practical reasons and may be the lumped source inductance and snubber inductance, that should have a freewheeling diode to suppress high voltages when the switch is opened.

Let us consider that steady conditions prevail. At the time t = 0– the switch Sw is open, the load current is $i_1 = I_1$, the diode current is $i_D = I_1 = I_F$, and the voltage drop V_D across the diode is small (about 1V).

The important concern is what happens after the switch is closed at t = 0. Figure 2–3(b) depicts the Waveforms of the diode current i_D and voltage v_D. At t = 0– there was the excess charge carrier distribution of conduction in the diode. This distribution cannot change instantaneously so at t = 0+ the diode still looks like a virtual short circuit, with $v_D = 1V$. Kirchhoff's current law provides us with the relation:

$$i_D = I_1 - i_s \quad (2.1)$$

and Kirchhoff's voltage law yields:

$$V_s = L_s \frac{di_s}{di_t} = L_s \frac{d(I_1 - i_D)}{dt} = -L_s \frac{di_D}{dt} \quad (2.2)$$

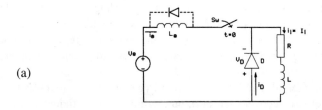

(a)

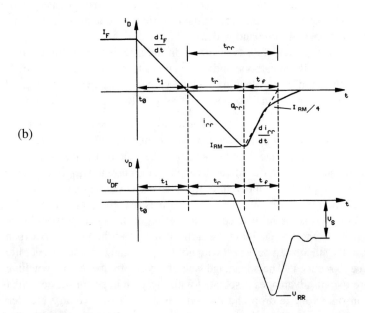

(b)

FIGURE 2-3 Diode turn-off (a) Chopper circuit (b) Waveforms

Accordingly, the diode current changes at the rate,

$$\frac{di_D}{dt} = -\frac{V_s}{L_s} = \text{Constant} \tag{2.3}$$

This means that it takes a time $t_1 = L_s \frac{I_1}{V_s}$ seconds for the diode current to fall to zero.

At the time $t = t_1$ the current i_D is zero, but up to this point the majority carriers have been crossing the junction to become minority carriers, so the P-N junction cannot assume a blocking condition until these carriers have been removed. At zero current, the diode is still a short circuit to the source voltage. Equations (2.1) to (2.3) still apply and the current i_D rises above I_l at the same rate. The diode voltage V_D changes little while the excess carriers remain. The diode reverse current rises over a time t_r during which the excess charge carriers are swept out of the region.

At the end of the interval t_r the reverse current i_D can have risen to a substantial value I_{RR} (peak reverse recovery current), but, by this time, sufficient carriers have been swept out and recombined that current cannot be supported. Therefore, over a fall-time interval t_f the diode current i_D reduces to almost zero very rapidly while the remaining excess carriers are swept out or recombined.

It is during the interval t_f that the potential barrier begins to increase both to block the reverse bias voltage applied by the source voltage as i_D reduces, and to suppress the diffusion of majority carriers because the excess carrier density at the junction is zero. The reverse voltage creates the electric field that allows the depletion layer to acquire space charge and widen.

That is, the electric field causes electrons in the n region to be forced away from the junction towards the cathode and causes holes to be forced away from the junction towards the anode. The blocking voltage v_D can rise above the voltage V_s transiently because of the additional voltage $L_s \frac{di_D}{dt}$ as i_D falls to zero over the time t_f.

The sum of the intervals $t_r + t_f = t_{rr}$ is known as the reverse recovery time and it varies generally (between 10 ns to over 1 microsecond) for different diodes. This time is also known as the storage time because it is the time that is taken to sweep out the excess charge Q_{RR} from the silicon by the reverse current. Q_{RR} is a function of $I_D = I_l$, $\frac{di_D}{dt}$ and the junction temperature. It has an effect on the reverse recovery current IRR and the reverse recovery time t_{rr}, so it is usually quoted in the data sheets. The fall time t_f can be influenced by the design of the diode. It would seem reasonable to make it short to decrease the turn-off time, but the process is expensive. The bulk of the silicon can be doped with gold or platinum to reduce carrier lifetimes and hence to reduce t_f. The advantage is an increased frequency of switching.

There are two disadvantages associated with this gain in performance. One is an increased on-state voltage drop and the other is an increased voltage recovery overshoot V_{RR}, that is caused by the increased $L_s \frac{di_s}{dt}$ as i_D falls more quickly.

Of the two effects, reverse recovery usually results in the greater power loss, and can also generate significant EMI. However these phenomena were considered to be no big deal at 50 or 60Hz. With the advent of semiconductor power switches, power conversion began to move into the multi-kilohertz range, and faster rectifiers were needed.

The relatively long minority carrier lifetime in silicon (tens of microseconds) causes a lot more charge to be stored than is necessary for effective conductivity modulation. In order to speed up reverse recovery, early "fast" rectifiers used various lifetime killing techniques to reduce the stored minority charge in the lightly doped region. The reverse recovery times of these rectifiers were dramatically reduced, down to about 200ns, although forward recovery and forward voltage were moderately increased as a side effect of the lifetime killing process. As power conversion frequencies increased to 20kHz and beyond, there eventually became a growing need for even faster rectifiers, which caused the "epitaxial" rectifier to be developed.

2.2.1.3 Fast and Ultra Fast Rectifiers

The foregoing discussion reveals the importance of the switching parameters such as (i) forward recovery time (t_{fr}), (ii) forward recovery voltage (V_{FR}), (iii) reverse recovery time (t_{rr}), (iv) reverse recovery charge (Q_{rr}), and (v) reverse recovery current I_{RM}, etc., during the transition from forward to reverse and vice versa. With various process improvements fast and ultrafast rectifiers have been achieved within the voltage and current limitations shown in Figure 2–1.

The figure shows that technology is available for devices up to 2000V ratings and over 1000 A current ratings which are mutually exclusive. In these diodes although cold t_{rr} values are good, at high junction temperature t_{rr} is three to four times higher, increasing switching losses and, in many cases, causing thermal runaway.

There exist several methods to control the switching characteristics of diodes and each leads to a different interdependency of forward voltage drop V_F, blocking voltage V_{RRM} and t_{rr} values. It is these interdependencies (or compromises) that differentiate the ultrafast diodes available on the market today. The important parameters for the turn-on and turn-off behavior of a diode are V_{FR}, V_F, t_{fr}, I_{RM} and t_{rr} and the values vary depending on the manufacturing processes.

Several manufacturers such as IXYS Semiconductors, International Rectifier, etc. manufacture a series of ultrafast diodes, termed Fast Recovery Epitaxial Diodes (FRED), which has gained wide acceptance during the 1990s. For an excellent description of these components see Burkel and Schneider (1994).

2.2.1.4 Schottky Rectifiers

Schottky rectifiers occupy a small corner of the total spectrum of available rectifier voltage and current ratings illustrated in Figure 2–1(b). They are, nonetheless, the rectifier of choice for low voltage switching power supply applications, with output voltages up to a few tens of volts, particularly at high switching frequency. For this reason, Schottkys account for a major segment of today's total rectifier usage. The Schottkys' unique electrical characteristics set them apart from conventional P-N junction rectifiers, in the following important respects:

- Lower forward voltage drop
- Lower blocking voltage
- Higher leakage current
- Virtual absence of reverse recovery charge

The two fundamental characteristics of the Schottky that make it a winner over the P-N junction rectifier in low voltage switching power supplies are its lower forward voltage drop, and virtual absence of minority carrier reverse recovery.

The absence of minority carrier reverse recovery means virtual absence of switching losses within the Schottky itself. Perhaps more significantly, the problem of switching voltage transients and attendant oscillations is less severe for Schottkys than for P-N junction rectifiers. Snubbers are therefore smaller and less dissipative.

The lower forward voltage drop of the Schottky means lower rectification losses, better efficiency, and smaller heat sinks. Forward voltage drop is a function of the Schottky's reverse voltage rating. The maximum voltage rating of today's Schottky rectifiers is about 150V. At this voltage, the Schottky's forward voltage drop is lower than that of a fast recovery epitaxial P-N junction rectifier by 150 to 200mV.

At lower voltage ratings, the lower forward voltage drop of the Schottky becomes progressively more pronounced, and more of an advantage. A 45V Schottky, for example, has a forward voltage drop of 0.4 to 0.6V, versus 0.85 to 1.0 V for a fast epitaxial P-N junction rectifier. A 15V Schottky has a mere 0.3 to 0.4V forward voltage drop.

A conventional fast recovery epitaxial P-N junction rectifier, with a forward voltage drop of 0.9V would dissipate about 18 percent of the output power of a 5V supply. A Schottky, by contrast, reduces rectification losses to the range of 8 to 12 percent. These are the simple reasons why Schottkys are virtually always preferred in low voltage high frequency switching power supplies. For any given current density, the Schottky's forward voltage drop increases as its reverse repetitive maximum voltage (V_{RRM}) increases. The basic hallmarks of any process are its maximum rated junction temperature—the T_{jmax} Class and the "prime" rated voltage, the V_{rrm} Class. These two basic hallmarks are set by the process; they in turn determine the forward voltage drop and reverse leakage current characteristics. Figure 2–4 indicates this condition for T_{jmax} of 150°c.

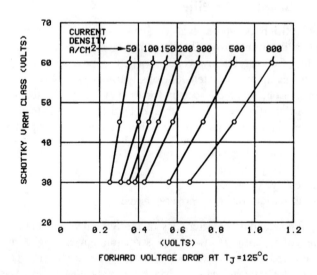

FIGURE 2–4 Relationships between Schottky V_{RRM} Class and forward voltage drop, for 150°c T_{jmax} class devices (Reproduced by permission of International Rectifier, USA)

2.2.1.4.1 Leakage Current and Junction Capacitance of Schottky Diodes

Figure 2–5 shows the dependence of leakage current on the operating voltage and junction temperature within any given process. Reverse leakage current increases with applied reverse voltage, and with junction temperature. Figure 2–6 shows typical relationship between operating temperature and leakage current, at rated V_{RRM}, for the 150°C/45V and 175°C/45V Schottky processes.

An important circuit-characteristic of the Schottky is its junction capacitance. This is a function of the area and thickness of the Schottky die, and of the applied voltage. The higher the V_{RRM} class, the greater the die thickness and the lower the junction capacitance. This is illustrated in Figure 2–7. Junction capacitance is essentially independent of the Schottky's T_{jmax} Class, and of operating temperature.

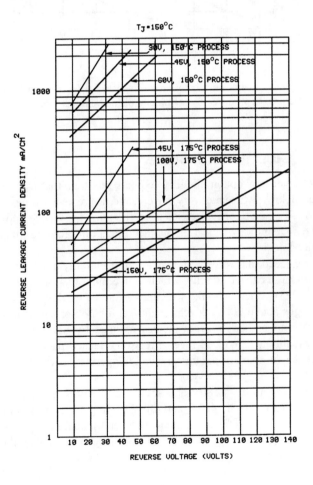

FIGURE 2–5 Relationships between reverse leakage current density, and applied reverse voltage (Reproduced by permission of International Rectifier, USA)

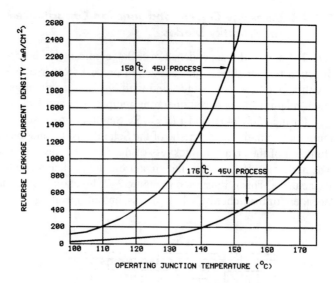

FIGURE 2-6 Typical relationships between reverse leakage current density, and operating junction temperature (Reproduced by permission of International Rectifier, USA)

2.2.1.5 GaAs Power Diodes

Efficient power conversion circuitry requires rectifiers that exhibit low forward voltage drop, low reverse recovery current, and fast recovery time. Silicon has been the material of choice for fast, efficient rectification in switched power applications. However, technology is nearing the theoretical limit for optimizing reverse recovery in silicon devices.

To increase speed, materials with faster carrier mobility are needed. Gallium Arsenide (GaAs) has a carrier mobility which is five times that of silicon (Delaney, Salih, and Lee 1995). Since Schottky technology for silicon devices is difficult to produce at voltages above 200V, development has focused on GaAs devices with ratings of 180V and higher. The advantages realized by using GaAs rectifiers include fast switching and reduced reverse recovery related parameters. An additional benefit is the variation of parameters with temperature is much less than silicon rectifiers.

For example, Motorola's 180V and 250V GaAs rectifiers are being used in power converters that produce 24, 36, and 48V DC outputs. Converters producing 48V DC, specially popular in telecommunications and mainframe computer applications, could gain the advantage of GaAs parts compared to similar silicon based parts at switching frequencies around 1MHz (Deuty 1996).

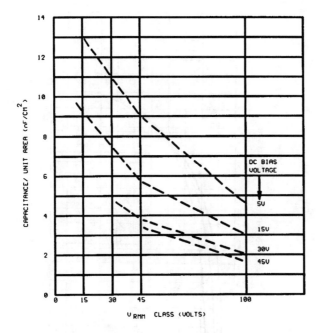

FIGURE 2–7 Typical Schottky self-capacitance versus V_{RRM} class, measured at various bias voltages (Reproduced by permission of International Rectifier, USA)

The 180V devices offered by Motorola can increase power density in 48V DC applications up to 90W/in^3 (21). These devices allow designers to switch converters at 1MHz without generating large amounts of EMI.

Figure 2–8(a) and 2–8(b) indicate typical forward voltage and typical reverse current for 20A, 180V, GaAs parts from Motorola.

For further details, the reader is directed to the following references: (Ashkanazi, Lorch and Nathan 1995), (Delaney, Salih, and Lee 1995), and (Deuty 1996).

2.2.2 Thyristors

The thyristor is a four-layer, three-terminal device as depicted in Figure 2–9. The complex interactions between three internal P-N junctions are then responsible for the device characteristics. However, the operation of the thyristor and the effect of the gate in controlling turn-on can be illustrated and followed by reference to the two transistor model of Figure 2–10. Here, the p_1-n_1-p_2 layers are seen to make up a p-n-p transistor and the n_2-p_2-n_1 layers create a n-p-n transistor with the collector of each transistor connected to the base of the other.

20 POWER ELECTRONICS DESIGN HANDBOOK

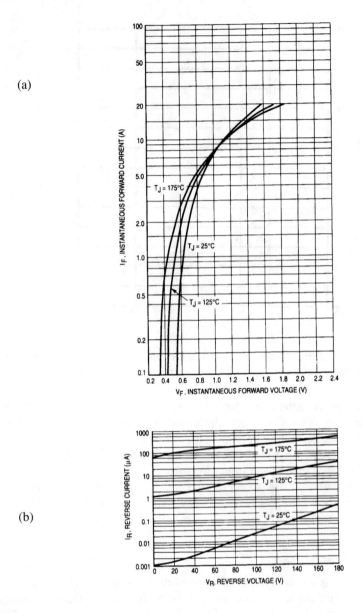

FIGURE 2-8 Typical characteristics of GaAs power diodes with 20A, 180V ratings (a) Forward voltage (b) Reverse current (Copyright of Motorola, used by permission)

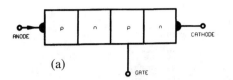
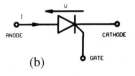

FIGURE 2-9 The thyristor (a) Construction (b) Circuit symbol

With a reverse voltage, cathode positive with respect to the anode, applied to the thyristor the p_1-n_1 and p_2-n_2 junctions are reverse biased and the resulting characteristic is similar to that of the diode with a small reverse leakage current flowing up to the point of reverse breakdown as shown by Figure 2–11(a). With a forward voltage, applied and no gate current supported the thyristor is in the forward blocking mode. The emitters of the two transistors are now forward biased and no conduction occurs. As the applied voltage is increased, the leakage current through the transistors increases to the point at which the positive feedback resulting from the base/collector connections drives both transistors into saturation, turning them, and hence the thyristor, on. The thyristor is now conducting and the forward voltage drop across it falls to a value of the order of 1 to 2 V. This condition is also shown in the thyristor static characteristic of Figure 2–11(a).

If a current is injected into the gate at a voltage below the breakover voltage, this will cause the n-p-n transistor to turn on. The positive feedback loop will then initiate the turn on of the p-n-p transistor. Once both transistors are on, the gate current can be removed because the action of the positive feedback loop will be to hold both transistors, and hence the thyristor, in the on state.

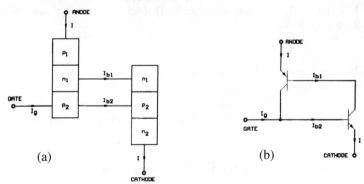

FIGURE 2-10 The two transistor model of a thyristor (a) Structure (b) The p-n-p and n-p-n transistor combination

22 POWER ELECTRONICS DESIGN HANDBOOK

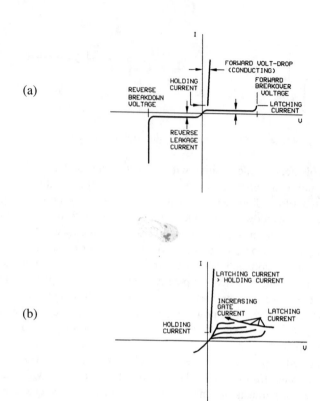

FIGURE 2-11 Thyristor characteristics (a) Thyistor characteristics with zero gate current (b) Switching characteristics

The effect of the gate current is therefore to reduce the effective voltage at which forward breakover occurs, as illustrated by the Figure 2–11(b). After the thyristor has been turned on it will continue to conduct as long as the forward current remains above the holding current level, irrespective of gate current or circuit conditions.

2.2.2.1 Ratings and Different Types of Devices

The operation of all power semiconductors is limited by a series of ratings which define the operating boundaries of the device. These ratings include limits on the peak, average and RMS currents, the peak forward and reverse voltages for the devices, maximum rates of change of device current and voltage, device junction temperature and, in the case of the thyristor, the gate current limits.

The current ratings of a power semiconductor are related to the energy dissipation in the device and hence the device junction temperature. The maximum value of on-state current ($I_{av(max)}$) is the maximum continuous current the device can sustain under defined conditions of voltage and current Waveform without exceeding the permitted temperature rise in the device. The rms current rating (I_{RMS}) is similarly related to the permitted temperature rise when operating into a regular duty cycle load.

In the case of transient loads, as the internal losses and hence the temperature rise in a power semiconductor are related to the square of the device forward current, the relationship between the current and the permitted temperature rise can be defined in terms of an $i^2 dt$ rating for the device. On turn-on, current is initially concentrated into a very small area of the device cross-section and the device is therefore subject to a di/dt rating which sets a limit to the permitted rate of rise of forward current.

The voltage ratings of a power semiconductor device are primarily related to the maximum forward and reverse voltages that the device can sustain. Typically, values will be given for the maximum continuous reverse voltage ($V_{RC(max)}$), the maximum repetitive reverse voltage ($V_{RR(max)}$) and the maximum transient reverse voltage ($V_{RT(max)}$). Similar values exist for the forward voltage ratings.

The presence of a fast transient of forward voltage can cause a thyristor to turn on and a dv/dt rating is therefore specified for the device. The magnitude of the imposed dv/dt can be controlled by the use of a snubber circuit connected in parallel with the thyristor. Data sheets for thyristors always quote a figure for the maximum surge current I_{TSM} that the device can survive.

This figure assumes a half sine pulse with a width of either 8.3 or 10 msec, which are the conditions applicable for 60 or 50Hz mains respectively. This limit is not absolute; narrow pulses with much higher peaks can be handled without damage, but little information is available to enable the designer to determine a current rating for short pulses. Hammerton (1989) indicates guidelines in this area.

Ever since its introduction, circuit design engineers have been subjecting the thyristor to increasing levels of operating stress and demanding that these devices perform satisfactorily there. The different stress demands that the thyristor must be able to meet are:

(a) Higher blocking voltages
(b) More current carrying capability
(c) Higher di/dt's
(d) Higher dv/dt's
(e) Shorter turn-off times
(f) Lower gate drive
(g) Higher operating frequencies.

There are many different thyristors available today which can meet one or more of these requirements, but as always, an improvement in one characteristic is usually only gained at the expense of another. As a result, different thyristors have been optimized for different applications. Modern thyristors can be classified into several general types, namely:

(a) Phase Control Thyristors
(b) Inverter Thyristors

(c) Asymmetrical Thyristors
(d) Reverse Conducting Thyristors (RCT)
(e) Light-Triggered Thyristors.

The voltage and current capabilities of phase control thyristors and inverter thyristors from a power device manufacturer are summarized in Figure 2–12.

2.2.2.1.1 Phase Control Thyristors

"Phase Control" or "Converter" thyristors generally operate at line frequency. They are turned off by natural commutation and do not have special fast-switching characteristics.

Current ratings of phase-control thyristors cover the range from a few amperes to about 3500A, and the voltage ratings from 50 to over 6500V. To simplify the gate-drive requirement and increase sensitivity, the use of amplifying gate, which was originally developed for fast switching "inverter" thyristors, is widely adopted in phase control SCR.

2.2.2.1.2 Inverter Thyristors

The most common feature of an inverter thyristor which distinguishes it from a standard phase control type is that it has fast turnoff time, generally in the range of 5 to 50 μs, depending upon voltage rating. Maximum average current ratings of over 2000 and 1300 A have been achieved with 2000 and 3000 V rated inverter thyristors, respectively.

Inverter thyristors are generally used in circuits that operate from DC supplies, where current in the thyristor is turned off either through the use of auxiliary comutating circuitry, by circuit resonance, or by "load" commutation. Whatever the circuit turn-off mechanism, fast turn-off is important because it minimizes sizes and weight of comutating and/or reactive circuit components.

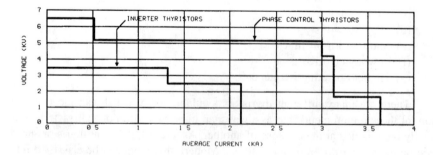

FIGURE 2-12 Thyristor rating capabilities (Reproduced with permission by GEC Plessey Semiconductors, UK)

2.2.2.1.3 Asymmetrical Thyristors

One of the main salient characteristics of asymmetrical thyristors (ASCR) is that they do not block significant reverse voltage. They are typically designed to have a reverse blocking capability in the range of 400 to 2000V.

The ASCR finds applications in many voltage-fed inverter circuits that require anti-parallel feedback rectifiers that keep the reverse voltage to less than 20V. The fact that ASCR needs only to block voltage in the forward direction provides an extra degree of freedom in optimizing turn-off time, turn-on time, and forward voltage drop.

2.2.2.1.4 Reverse Conducting Thyristors

The reverse conducting thyristor (RCT) represents the monolithic integration of an asymmetrical thyristor with an anti-parallel rectifier. Beyond obvious advantages of the parts count reduction, the RCT eliminates the inductively induced voltage within the thyristor-diode loop (virtually unavoidable to some extent with separate discrete components). Also, it essentially limits the reverse voltage seen by the thyristor to only the conduction voltage of the diode.

2.2.2.1.5 Light-Triggered Thyristors

Many developments have taken place in the area of light-triggered thyristors. Direct irradiation of silicon with light created electron-hole pairs which, under the influence of an electric field, produce a current that triggers the thyristors.

The turn-on of a thyristor by optical means is an especially attractive approach for devices that are to be used in extremely high-voltage circuits. A typical application area is in switches for DC transmission lines operating in the hundreds of kilovolts range, which use series connections of many devices, each of which must be triggered on command. Optical firing in this application is ideal for providing the electrical isolation between trigger circuits and the thyristor which floats at a potential as high as hundreds of kilovolts above ground.

The main requirement for an optically-triggered thyristor is high sensitivity while maintaining high dv/dt and di/dt capabilities. Because of the small and limited quantity of photo energy available for triggering the thyristor from practical light sources, very high gate sensitivity of the order of 100 times that of the electrically triggered device is needed.

2.2.2.1.6 JEDEC Titles and Popular Names

Table 2-1 compares the Joint Electronic Device Engineering Council (JEDEC) titles for commercially available thyristors types with popular names. JEDEC is an industry standardization activity co-sponsored by the Electronic Industries Association (EIA) and the National Manufacturers Association (NEMA). Silicon controlled rectifiers (SCR) are the most widely used as power control elements. Triacs are quite popular in lower current (<40A) AC power applications.

TABLE 2-1   Thyristor Types and Popular Names

JEDEC Titles	Popular Names, Types
Reverse Blocking Diode Thyristor	* Four Layer Diode, Silicon Unilateral Switch (SUS)
Reverse Blocking Triode Thyristor	Silicon Controlled Rectifier (SCR)
Reverse Conducting Diode Thyristor	* Reverse Conducting Four Layer Diode
Reverse Conducting Triode Thyristor	Reverse Conducting SCR
Bidirectional Triode Thyristor	Triac
Turn-Off Thyristor	Gate Turn Off Switch (GTO)

* Not generally available

2.3   Gate Turn-Off Thyristors

A gate turn-off thyristor (GTO) is a thyristor-like latching device that can be turned off by application of a negative pulse of current to its gate. This gate turn-off capability is advantageous because it provides increased flexibility in circuit application. It now becomes possible to control power in DC circuits without the use of elaborate commutation circuitry.

Prime design objectives for GTO devices are to achieve fast turn-off time and high current turn-off capability and to enhance the safe operating area during turn-off. Significant progress has been made in both areas during the last few years, largely due to a better understanding of the turn-off mechanisms. The GTO's turn-off occurs by removal of excess holes in the cathode-base region by reversing the current through the gate terminal.

The GTO is gaining popularity in switching circuits, especially in equipment which operates directly from European mains. The GTO offers the following advantages over a bipolar transistor: high blocking voltage capabilities, in excess of 1500V, and also high over-current capabilities. It also exhibits low gate currents, fast and efficient turn-off, as well as outstanding static and dynamic dv/dt capabilities.

Figure 2–13(a) depicts the symbol of GTO and Figure 2–13(b) shows its two transistor equivalent circuit. Figure 2–13(c) shows a basic drive circuit. The GTO is turned on by a positive gate current, and it is turned off by applying a negative gate cathode voltage.

A practical implementation of a GTO gate drive circuit is shown in Figure 2–14. In this circuit when transistor Q_2 is off, emitter follower transistor Q_1 acts as a current source pumping current into the gate of the GTO through a 12-V Zener Z_1 and polarized capacitor C_1. When the control voltage at the base of Q_2 goes positive, transistor Q_2 turns on, while transistor Q_1 turns off since its base now is one diode drop more negative than its emitter. At this stage the positive side of capacitor C_1 is essentially grounded, and C_1 will act as a voltage source of approximately

Power Semiconductors   27

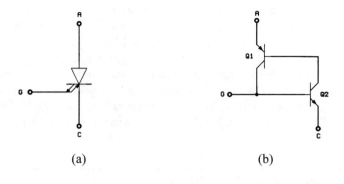

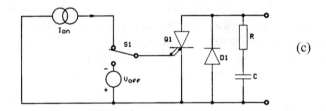

FIGURE 2-13 GTO symbol, equivalent circuit & basic drive circuit (a) Symbol of GTO (b) Two transistor equivalent of GTO (c) Basic drive circuit

10V, turning the GTO off. Isolated gate drive circuits may also be easily implemented to drive the GTO.

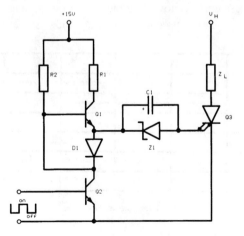

FIGURE 2-14 Practical realization of a GTO gate drive circuit.

28 POWER ELECTRONICS DESIGN HANDBOOK

With improved cathode emitter geometries and better optimized vertical structures, today's GTOs have made significant progress in turn-off performance (the prime weakness of earlier day GTOs). Figure 2–15 shows the available GTO ratings and, as can be seen, they cover quite a wide spectrum. However, the main applications lie in the higher voltage end (>1200 V) where bipolar transistors and power MOSFETs are unable to compete effectively. In the present day market there are GTOs with current ratings over 3000A and voltage ratings over 4500V. For further details on GTOs see Coulbeck, Findlay, and Millington (1994) and Bassett and Smith (1989).

2.4 Bipolar Power Transistors

During the last two decades, attention has been focused on high power transistors as switching devices in inverters, SMPS and similar switching applications. New devices with faster switching speeds and lower switching losses are being developed that offer performance beyond that of thyristors. With their faster speed, they can be used in an inverter circuit operating at frequencies over 200 KHz. In addition, these devices can be readily turned off with a low-cost reverse base drive without the costly commutation circuits required by thyristors.

2.4.1 Bipolar Transistor as a Switch

The bipolar transistor is essentially a current-driven device. That is, by injecting a current into the base terminal a flow of current is produced in the collector. There are essentially two modes of operation in a bipolar transistor: the linear and saturating modes. The linear mode is used when amplification is needed, while the saturating mode is used to switch the transistor either on or off.

Figure 2–16 shows the V-I characteristic of a typical bipolar transistor. Close examination of these curves shows that the saturation region of the V-I curve is of interest when the transistor is used in a switching mode. At that region a certain base

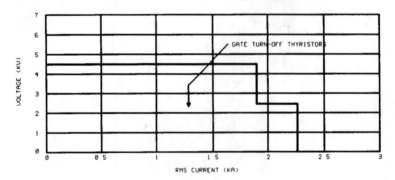

FIGURE 2-15 Ratings covered by available GTOs (Reproduced by permission of GEC Plessey Semiconductors, UK)

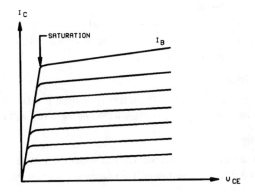

FIGURE 2-16   Typical output characteristics of BJT

current can switch the transistor on, allowing a large amount of collector current to flow, while the collector-to-emitter voltage remains relatively small.

In actual switching applications a base drive current is needed to turn the transistor on, while a base current of reverse polarity is needed to switch the transistor back off. In practical switching operations certain delays and storage times are associated with transistors. In the following section are some parameter definitions for a discrete bipolar transistor driven by a step function into a resistive load.

Figure 2-17 illustrates the base-to-emitter and collector-to-emitter waveforms of a bipolar NPN transistor driven into a resistive load by a base current pulse I_B. The following are the definitions associated with these waveforms:

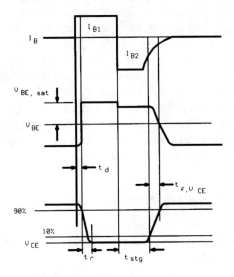

FIGURE 2-17   Bipolar transistor switching waveforms

Delay Time, t_d

Delay time is defined as the interval of time from the application of the base drive current I_{B1} to the point at which the collector-emitter voltage V_{CE} has dropped to 90 percent of its initial off value.

Rise Time, t_r

Rise time is defined as the interval of time it takes the collector-emitter voltage V_{CE} to drop to 10 percent from its 90 percent off value.

Storage Time, t_{stg}

Storage time is the interval of time from the moment reverse base drive I_{B2} is applied to the point where the collector-emitter voltage V_{CE} has reached 10 percent of its final off value.

Fall Time, $t_{f,VCE}$

Fall time is the time interval required for the collector-emitter voltage to increase from 10 to 90 percent of its off value.

2.4.2 Inductive Load Switching

In the previous section, the definitions for the switching times of the bipolar transistor were made in terms of collector-emitter voltage. Since the load was defined to be a resistive one, the same definitions hold true for the collector current. However, when the transistor drives an inductive load, the collector voltage and current waveforms will differ. Since current through an inductor does not flow instantaneously with applied voltage, during turn-off, one expects to see the collector-emitter voltage of a transistor rise to the supply voltage before the current begins to fall. Thus, two fall time components may be defined, one for the collector-emitter voltage $t_{f,VCE}$, and the other for the collector current $t_{f,Ic}$. Figure 2–18 shows the actual waveforms.

Observing the waveforms we can define the collector-emitter fall time $t_{f,VCE}$ in the same manner as in the resistive case, while the collector fall time $t_{f,Ic}$ may be defined as the interval in which collector current drops from 90 to 10 percent of its initial value. Normally, the load inductance L behaves as a current source, and therefore it charges the base-collector transition capacitance faster than the resistive load. Thus, for the same base and collector currents the collector-emitter voltage fall time $t_{f,VCE}$ is shorter for the inductive circuit.

2.4.3 Safe Operating Area and V-I Characteristics

The output characteristics (I_C Versus V_{CE}) of a typical npn power transistor are shown in Figure 2–19(a). The various curves are distinguished from each other by the value of the base current.

Several features of the characteristics should be noted. First, there is a maximum collector-emitter voltage that can be sustained across the transistor when it is

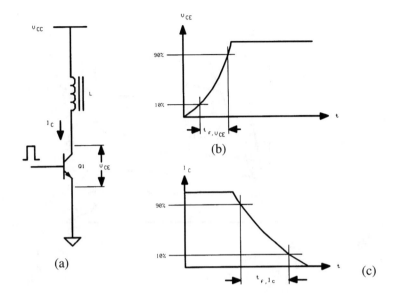

FIGURE 2-18 A bipolar switching transistor driving an inductive load with associated fall time waveforms (a) Circuit (b) Voltage waveform (c) Current waveform

carrying substantial collector current. The voltage is usually labeled BV_{SUS}. In the limit of zero base current, the maximum voltage between collector and emitter that can be sustained increases somewhat to a value labeled BV_{CEO}, the collector-emitter breakdown voltage when the base is open circuited. This latter voltage is often used as the measure of the transistor's voltage standoff capability because usually the only time the transistor will see large voltages is when the base current is zero and the BJT is in cutoff.

The voltage BV_{CBO} is the collector-base breakdown voltage when the emitter is open circuited. The fact that this voltage is larger than BV_{CEO} is used to advantage in so-called open-emitter transistor turn-off circuits.

The region labeled primary breakdown is due to conventional avalanche breakdown of the collector-base junction and the attendant large flow of current. This region of the characteristics is to be avoided because of the large power dissipation that clearly accompanies such breakdown.

The region labeled second breakdown must also be avoided because large power dissipation also accompanies it, particularly at localized sites within the semiconductor. The origin of second breakdown is different from that of avalanche breakdown and will be considered in detail later in this chapter. BJT failure is often associated with second breakdown.

The major observable difference between the I-V characteristics of a power transistor and those of a logic level transistor is the region labeled quasi-saturation on the power transistor characteristics of Figure 2–19(a). Quasi-saturation is a consequence of the lightly doped collector drift region found in the power transistor.

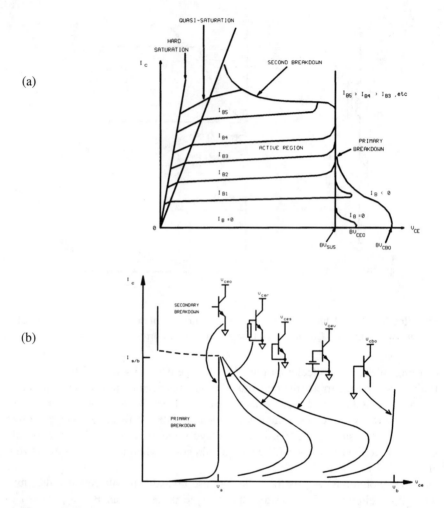

FIGURE 2-19 Current-voltage characteristics of a NPN power transistor showing breakdown phenomenon (a) indication of quasi saturation (b) relative primary and secondary breakdown conditions for different bias levels

Logic level transistors do not have this drift region and so do not exhibit quasi-saturation. Otherwise all of the major features of the power transistor characteristic are also found on those of logic level devices.

Figure 2–19(b) indicates the relative magnitudes of npn transistor collector breakdown characteristics, showing primary and secondary breakdown with different base bias conditions. With low gain devices V_a is close to V_b in value, but with high gain devices V_b may be 2 to 3 times that of V_a. Notice that negative resistance characteristics occur after breakdown, as is the case with all the circuit-dependent breakdown characteristics. (B. W. Williams 1992) provides a detailed explanation on the behavior.

2.4.3.1 Forward-Bias Secondary Breakdown

In switching process **BJT**s are subjected to great stress, during both turn-on and turn-off. It is imperative that the engineer clearly understands how the power bipolar transistor behaves during forward and reverse bias periods in order to design reliable and trouble-free circuits.

The first problem is to avoid secondary breakdown of the switching transistor at turn-on, when the transistor is forward-biased. Normally the manufacturer's specifications will provide a safe-operating area (SOA) curve, such as the typical one shown in Figure 2–20. In this figure collector current is plotted against collector-emitter voltage. The curve locus represents the maximum limits at which the transistor may be operated. Load lines that fall within the pulsed forward-bias SOA curve during turn-on are considered safe, provided that the device thermal limitations and the SOA turn-on time are not exceeded.

The phenomenon of forward-biased secondary breakdown is caused by hot spots which are developed at random points over the working area of a power transistor, caused by unequal current conduction under high-voltage stress. Since the temperature coefficient of the base-to-emitter junction is negative, hot spots increase local current flow. More current means more power generation, which in turn raises the temperature of the hot spot even more.

Since the temperature coefficient of the collector-to-emitter breakdown voltage is also negative, the same rules apply. Thus is the voltage stress is not removed, ending the current flow, the collector-emitter junction breaks down and the transistor fails because of thermal runaway.

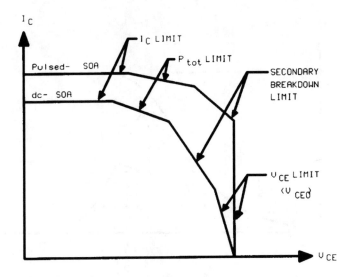

FIGURE 2–20 DC and pulse SOA for BJT

2.4.3.2 Reverse-Bias Secondary Breakdown

It was mentioned in previous paragraphs that when a power transistor is used in switching applications, the storage time and switching losses are the two most important parameters with which the designer has to deal extensively.

On the other hand the switching losses must also be controlled since they affect the overall efficiency of the system. Figure 2–21 shows turn-off characteristics of a high-voltage power transistor in resistive and inductive loads. Inspecting the curves we can see that the inductive load generates a much higher peak energy at turn-off than its resistive counterpart. It is then possible, under these conditions, to have a secondary breakdown failure if the reverse-bias safe operating area (RBSOA) is exceeded.

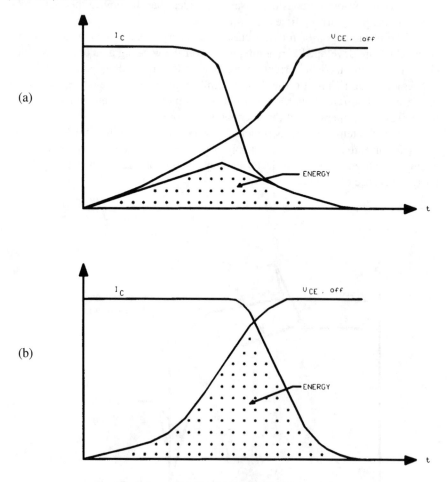

FIGURE 2-21 Turn-off characteristics of a high voltage BJT (a) Resistive load (b) Inductive load

The RBSOA curve (Figure 2–22) shows that for voltages below V_{CEO} the safe area is independent of reverse-bias voltage V_{EB} and is only limited by the device collector current I_C. Above V_{CEO} the collector current must be derated depending upon the applied reverse-bias voltage.

It is then apparent that the reverse-bias voltage V_{EB} is of great importance and its effect on RBSOA very interesting. It is also important to remember that avalanching the base-emitter junction at turn-off must be avoided, since turn-off switching times may be decreased under such conditions. In any case, avalanching the base-emitter junction may not be considered relevant, since normally designers protect the switching transistors with either clamp diodes or snubber networks to avoid such encounters.

2.4.4 Darlington Transistors

The on-state voltage $V_{CE(sat)}$ of the power transistors is usually in the 1–2 V range, so that the conduction power loss in the BJT is quite small. BJTs are current-controlled devices and base current must be supplied continuously to keep them in the on-state. The DC current gain h_{FE} is usually only 5–10 in high-power transistors and so these devices are sometimes connected in a Darlington or triple Darlington configuration as is shown in Figure 2–23 to achieve a larger current gain. However, some disadvantages accrue in this configuration including slightly higher overall $V_{CE(sat)}$ values and slower switching speeds. The current gain of the pair h_{FE} is

$$h_{FE} = h_{FE1} \cdot h_{FE2} + h_{FE1} + h_{FE2} \qquad (2.4)$$

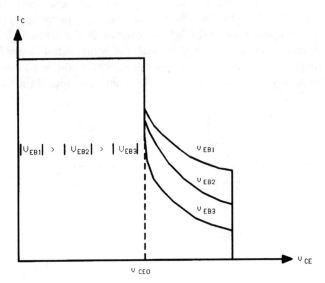

FIGURE 2–22 RBSOA plot for a high voltage BJT as a function of reverse-bias voltage V_{EB}

Darlington configurations using discrete BJTs or several transistors on a single chip (a Monolithic Darlington (MD)) have significant storage time during the turn-off transition. Typical switching times are in the range of a few hundred nanoseconds to a few microseconds.

BJTs including MDs are available in voltage ratings up to 1400 V and current ratings of a few hundred amperes. In spite of a negative temperature coefficient of on-state resistance, modern BJTs fabricated with good quality control can be paralleled provided that care is taken in the circuit layout and that some extra current margin is provided.

Figure 2–24 shows a practical Monolithic Darlington pair with diode D_1 added to speed up the turn-off time of Q_1 and D_2 added for half and full bridge circuit applications. Resistors R_1 and R_2 are low value resistors and provide a leakage path for Q_1 and Q_2.

2.5 Power MOSFETs

2.5.1 Introduction

Compared to BJTs which are current controlled devices, field effect transistors are voltage controlled devices. There are two basic field-effect transistors (FETs): the junction FET (JFET) and the metal-oxide semiconductor FET (MOSFET). Both have played important roles in modern electronics. The JFET has found wide application in such cases as high-impedance transducers (scope probes, smoke detectors, etc.) and the MOSFET in an ever-expanding role in integrated circuits where CMOS (Complementary MOS) is perhaps the most well-known.

Power MOSFETs differ from bipolar transistors in operating principles, specifications, and performance. In fact, the performance characteristics of MOSFETS are generally superior to those of bipolar transistors: significantly faster switching time, simpler drive circuitry, the absence of a second-breakdown failure mechanism, the ability to be paralleled, and stable gain and response time over a wide tem-

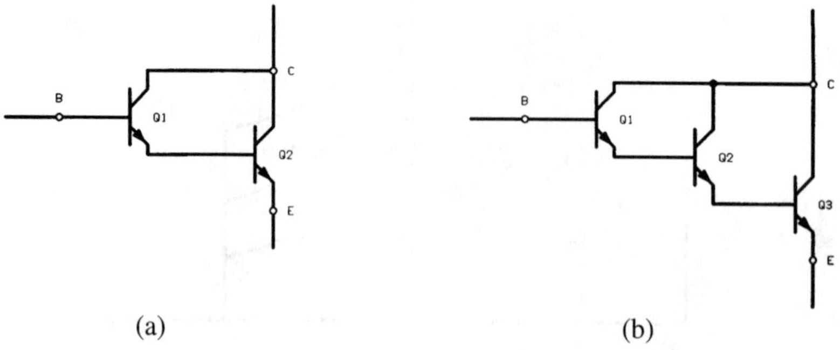

(a)   (b)

FIGURE 2–23   Darlington configurations (a) Darlington (b) Triple darlington

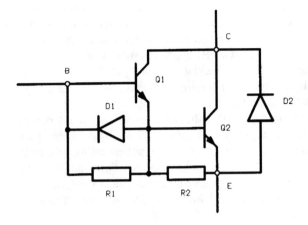

FIGURE 2-24 A practical Monolithic Darlington pair

perature range. The MOSFET was developed out of the need for a power device that could work beyond the 20 KHz frequency spectrum, anywhere from 100 KHz to above 1 MHz, without experiencing the limitations of the bipolar power transistor.

2.5.2 General Characteristics

Bipolar transistors are described as minority-carrier devices in which injected minority carriers recombine with majority carriers. A drawback of recombination is that it limits the device's operating speed. Current-driven base-emitter input of a bipolar transistor presents a low-impedance load to its driving circuit. In most power circuits, this low-impedance input requires somewhat complex drive circuitry.

By contrast, a power MOSFET is a voltage-driven device whose gate terminal is electrically isolated from its silicon body by a thin layer of silicon dioxide (SiO_2). As a majority-carrier semiconductor, the MOSFET operates at much higher speed than its bipolar counterpart because there is no charge-storage mechanism. A positive voltage applied to the gate of an n-type MOSFET creates an electric field in the channel region beneath the gate; that is, the electric charge on the gate causes the p-region beneath the gate to convert to an n-type region, as shown in Figure 2–25(a).

This conversion, called the surface-inversion phenomenon, allows current to flow between the drain and source through an n-type material. In effect, the MOSFET ceases to be an n-p-n device when in this state. The region between the drain and source can be represented as a resistor, although it does not behave linearly, as a conventional resistor would. Because of this surface-inversion phenomenon, then, the operation of a MOSFET is entirely different from that of a bipolar transistor.

By virtue of its electrically-isolated gate, a MOSFET is described as a high-input impedance, voltage-controlled device, compared to a bipolar transistor. As a majority-carrier semiconductor, a MOSFET stores no charge, and so can switch faster than a bipolar device. Majority-carrier semiconductors also tend to slow down as temperature increases. This effect brought about by another phenomenon called carrier mobility makes a MOSFET more resistive at elevated temperatures, and much more immune to the thermal-runaway problem experienced by bipolar devices. Mobility is a term that defines the average velocity of a carrier in terms of the electrical field imposed on it.

A useful by-product of the MOSFET process is the internal parasitic diode formed between source and drain, Figure 2–25(b). (There is no equivalent for this diode in a bipolar transistor other than in a bipolar Darlington transistor.) Its characteristics make it useful as a clamp diode in inductive-load switching.

Different manufacturers use different techniques for constructing a power FET, and names like HEXFET, VMOS, TMOS, etc., have become trademarks of specific companies.

2.5.3 MOSFET Structures and On Resistance

Most Power MOSFETs are manufactured using various proprietary processes by various manufactures on a single silicon chip structured with a large number of closely packed identical cells. For example, Harris Power MOSFETs are manufactured using a vertical double-diffused process, called VDMOS or simply DMOS. In these cases, a 120-mil^2 chip contains about 5,000 cells and a 240-mil^2 chip has more than 25,000 cells.

One of the aims of multiple-cells construction is to minimize the MOSFET parameter $r_{DS(ON)}$ when the device is in the on-state. When $r_{DS(ON)}$ is minimized, the device provides superior power-switching performance because the voltage drop from drain to source is also minimized for a given value of drain-source current. Reference 6 provides more details.

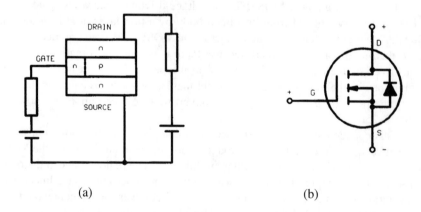

FIGURE 2–25 Structure of N-channel MOSFET and symbol (a) Structure (b) Symbol

2.5.4 I-V Characteristics

Figure 2–26 shows the drain-to-source operating characteristics of the power MOSFET. Although curve is similar to the case of bipolar power transistor (Figure 2–15), there are some fundamental differences.

The MOSFET output characteristic curves reveal two distinct operating regions, namely, a "constant resistance" and a "constant current." Thus, as the drain-to-source voltage is increased, the drain current increases proportionally, until a certain drain-to-source voltage called "pinch off" is reached. After pinch off, an increase in drain-to-source voltage produces a constant drain current.

When the power MOSFET is used as a switch, the voltage drop between the drain and source terminals is proportional to the drain current; that is, the power MOSFET is working in the constant resistance region, and therefore it behaves essentially as a resistive element. Consequently the on-resistance $R_{DS,on}$ of the power MOSFET is an important figure of merit because it determines the power loss for a given drain current, just as $V_{CE,sat}$ is of importance for the bipolar power transistor.

By examining Figure 2–26, we note that the drain current does not increase appreciably when a gate-to-source voltage is applied; in fact, drain current starts to flow after a threshold gate voltage has been applied, in practice somewhere between 2 and 4V. Beyond the threshold voltage, the relationship between drain current and gate voltage is approximately equal. Thus, the transconductance g_{fs}, which is defined as the rate of change of drain current to gate voltage, is practically constant at higher values of drain current. Figure 2–27 illustrates the transfer characteristics of I_D versus V_{DS}, while Figure 2–28 shows the relationship of transconductance g_{fs} to drain current.

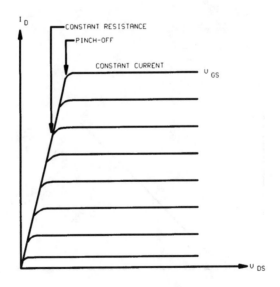

FIGURE 2–26   Typical output characteristics of a MOSFET

It is now apparent that a rise in transconductance results in a proportional rise in the transistor gain, i.e., larger drain current flow, but unfortunately this condition swells the MOSFET input capacitance. Therefore, carefully designed gate drivers must be used to deliver the current required to charge the input capacitance in order to enhance the switching speed of the MOSFET.

2.5.5 Gate Drive Considerations

The MOSFET is a voltage controlled device; that is, a voltage of specified limits must be applied between gate and source in order to produce a current flow in the drain.

Since the gate terminal of the MOSFET is electrically isolated from the source by a silicon oxide layer, only a small leakage current flows from the applied voltage source into the gate. Thus, the MOSFET has an extremely high gain and high impedance.

In order to turn a MOSFET on, a gate-to-source voltage pulse is needed to deliver sufficient current to charge the input capacitor in the desired time. The MOSFET input capacitance C_{iss} is the sum of the capacitors formed by the metal-oxide gate structure, from gate to drain (C_{GD}) and gate to source (C_{GS}). Thus, the driving voltage source impedance R_g must be very low in order to achieve high transistor speeds. A way of estimating the approximate driving generator impedance, plus the required driving current, is given in the following equations:

$$R_g = \frac{t_r (\text{or } t_f)}{2.2 C_{iss}} \quad (2.5)$$

and

$$I_g = C_{iss} \cdot \frac{dv}{dt} \quad (2.6)$$

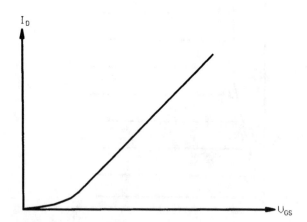

FIGURE 2-27　Transfer characteristics of a power MOSFET

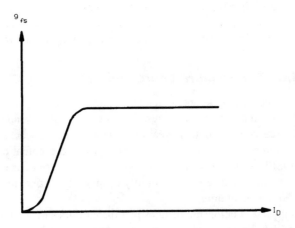

FIGURE 2-28 Relationship of transconductance (g_{fs}) to I_D of a power MOSFET

Where R_g = generator impedance,
C_{iss} = MOSFET input capacitance, pF
dv/dt = generator voltage rate of change, V/ns

To turn off the MOSFET, we need none of the elaborate reverse current generating circuits described for bipolar transistors. Since the MOSFET is a majority carrier semiconductor, it begins to turn off immediately upon removal of the gate-to-source voltage. Upon removal of the gate voltage the transistor shuts down, presenting a very high impedance between drain and source, thus inhibiting any current flow, except leakage currents (in microamperes).

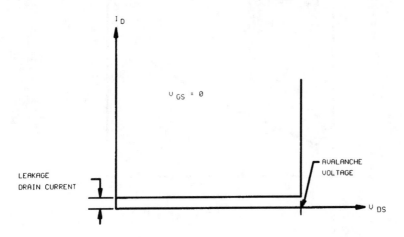

FIGURE 2-29 Drain-to-source blocking characteristics of the MOSFET

Figure 2–29 illustrates the relationship of drain current versus drain-to-source voltage. Note that drain current starts to flow only when the drain-to-source avalanche voltage is exceeded, while the gate-to-source voltage is kept at 0V.

2.5.6 Temperature Characteristics

The high operating temperatures of bipolar transistors are a frequent cause of failure. The high temperatures are caused by hot-spotting, the tendency of current in a bipolar device to concentrate in areas around the emitter. Unchecked, this hot spotting results in the mechanism of thermal runaway, and eventual destruction of the device. MOSFETs do not suffer this disadvantage because their current flow is in the form of majority carriers. The mobility of majority carriers in silicon decreases with increasing temperature.

This inverse relationship dictates that the carriers slow down as the chip gets hotter. In effect, the resistance of the silicon path is increased, which prevents the concentrations of current that lead to hot spots. In fact if hot spots do attempt to form in a MOSFET, the local resistance increases and defocuses or spreads out the current, rerouting it to cooler portions of the chip.

Because of the character of its current flow, a MOSFET has a positive temperature coefficient of resistance, as shown by the curves of Figure 2–30.

The positive temperature coefficient of resistance means that a MOSFET is inherently stable with temperature fluctuation, and provides its own protection against thermal runaway and second breakdown. Another benefit of this characteristic is that MOSFET can be operated in parallel without fear that one device will rob current from the others. If any device begins to overheat, its resistance will increase, and its current will be directed away to cooler chips.

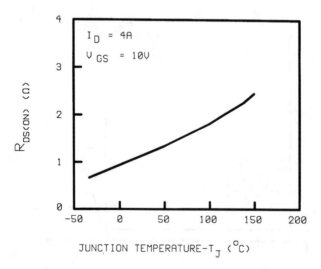

FIGURE 2–30 Positive temperature coefficient of a MOSFET

2.5.7 Safe Operating Area

In the discussion of the bipolar power transistor, it was mentioned that in order to avoid secondary breakdown, the power dissipation of the device must be kept within the operating limits specified by the forward-bias SOA curve. Thus, at high collector voltages the power dissipation of the bipolar transistor is limited by its secondary breakdown to a very small percentage of full rated power. Even at very short switching periods the SOA capability is still restricted, and the use of snubber networks is incorporated to relieve transistor switching stress and avoid secondary breakdown.

In contrast, the MOSFET offers an exceptionally stable SOA, since it does not suffer from the effects of secondary breakdown during forward bias. Thus, both the DC and pulsed SOA are superior to that of the bipolar transistor. In fact with a power MOSFET it is quite possible to switch rated current at rated voltage without the need of snubber networks. Of course, during the design of practical circuits, it is advisable that certain derating must be observed.

Figure 2–31 shows typical MOSFET and equivalent bipolar transistor curves superimposed in order to compare their SOA capabilities. Secondary breakdown during reverse bias is also nonexistent in the power MOSFET, since the harsh reverse-bias schemes used during bipolar transistor turn-off are not applicable to MOSFETs. Here, for the MOSFET to turn off, the only requirement is that the gate is returned to 0 V.

2.5.8 Practical Components

2.5.8.1 High Voltage and Low On Resistance Devices

Denser geometries, processing innovations, and packaging improvements are resulting in power MOSFETs that have ever-higher voltage ratings and current-handling capabilities, as well as volumetric power-handling efficiency. See Travis

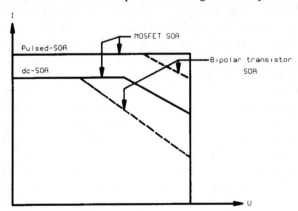

FIGURE 2–31 SOA curves for power MOSFET

(1989), Goodenough (1995), and Goodenough (1994) for more information. Bipolar transistors have always been available with very high voltage ratings, and those ratings do not carry onerous price penalties. Achieving good high-voltage performance in power MOSFETS, however, has been problematic, for several reasons.

First, the $r_{DS(on)}$ of devices of equal silicon area increases exponentially with the voltage rating. To get the on-resistance down, manufacturers would usually pack more parallel cells onto a die. But this denser packing causes problems in high-voltage performance. Propagation delays across a chip as well as silicon defects, can lead to unequal voltage stresses and even to localized breakdown.

Manufacturers resort to a variety of techniques to produce high voltage (> 1000V), low-$r_{DS(on)}$ power MOSFETS that offer reasonable yields. Advanced Power Technology (APT), for example, deviates from the trend toward smaller and smaller feature sizes in its quest for low on-resistance. Instead, the company uses large dies to get $r_{DS(on)}$ down. APT manufactures power MOSFETS using dies as large as 585 x 738 mil and reaching voltage ratings as high as 1000V. APT10026JN, a device from their product range has a current rating of 1000V with 690W power rating. On resistance of the device is 0.26Ω.

When devices are aimed out to low power applications such as laptop and notebook computers, personal digital assistants (PDA), etc. extremely low $R_{DS(on)}$ values from practical devices are necessary.

Specific on-resistance of double-diffused MOSFETs (DMOSFETs), more commonly known as power MOSFETs, has continually shrunk over the past two decades. In other words, the $R_{DS(on)}$ per unit area has dropped. The reduced size with regard to low-voltage devices (those rated for a maximum drain-to-source voltage, V_{DS} of under 100 V) was achieved by increasing the cell density.

Most power-MOSFET suppliers now offer low-voltage FETs from processes that pack 4,000,000 to 8,000,000 cells/in^2, in which each cell is an individual MOSFET. The drain, gate, and source terminals of all the cells are connected in parallel. Manufacturers such as International Rectifier (IOR) have developed many generations of MOSFETs based on DMOS technology. For example, the HEXFET family from IOR has gone through five generations, gradually increasing the number of cells per in^2 with almost 10 fold decrease in the $R_{DS(on)}$ parameter as per Figure 2–32. $R_{DS(on)}$ times the device die area in this figure is a long used figure of merit (FOM) for power semiconductors. This is called "specified on resistance." For details Kinzer (1995) is suggested.

Because generation 5 die are smaller than the previous generation, there is room within the same package to accommodate additional devices such as a Schottky diode. The FETKY family from IOR (Davis 1997) which uses this concept of integrating a MOSFET with a Schottky diode is aimed for power converter applications such as synchronous regulators, etc.

In designing their DMOSFETs, Siliconix borrowed a DRAM process technique called the trench gate. They then developed a low voltage DMOSFET process that provides 12,000,000 cells/in^2, and offers lower specific on-resistance— $R_{DS(on)}$—than present planar processes.

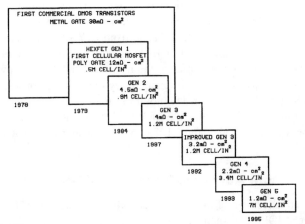

FIGURE 2-32 The Power MOSFETs Generations (Courtesy: International Rectifier)

The first device from the process, the n-channel Si4410DY, comes in Siliconix's data book "Little Foot" 8-pin DIP. The Si4410DY sports a maximum on-resistance of 13.5 m Ω, enhanced with 10V of gate-to-source voltage (V_{GS}). At a V_{GS} of 4.5 V $R_{DS(on)}$ nearly doubles, reaching a maximum value of 20 mΩ.

During the year 1997 Temic Semiconductors (formerly Siliconix) has further improved their devices to carry 32 million cells/in[2] using their "TrenchFET" technology. These devices come in two basic families namely (i) Low-on resistance devices and (ii) Low-threshold devices. Maximum on-resistance were reduced to 9mΩ and 13mΩ, compared to the case of Si 4410DY devices with corresponding values of 13.5mΩ (for V_{GS} of 10V) and 20mΩ (for V_{GS} of 4.5V), for these low on resistance devices. In the case of low-threshold family these values were 10 mΩ and 14 mΩ for gate source threshold values of 4.5V and 2.5V respectively. For details Goodenough (1997) is suggested.

2.5.8.2 P-Channel MOSFETS

Historically, p-channel FETs were not considered as useful as their n-channel counterparts. The higher resistivity of p-type silicon, resulting from its lower carrier mobility, put it at a disadvantage compared to n-type silicon.

Due to the approximately 2:1 superior mobility on n-type devices, n-channel power FETs dominate the available devices, because they need about half the area of silicon for a given current or voltage rating. However, as the technology matures, and with the demands of power-management applications, p-channel devices are starting to become available.

They make possible power CMOS designs and eliminate the need for special high-side drive circuits. When a typical n-channel FET is employed as a high-side switch running off a plus supply rail with its source driving the load, the gate must be pulled at least 10 V above the drain. A p-channel FET has no such requirement. A high-side p-channel MOSFET and a low-side n-channel MOSFET tied with common drains make a superb high-current "CMOS equivalent" switch.

46 POWER ELECTRONICS DESIGN HANDBOOK

Because on-resistance rises rapidly with device voltage rating, it was only recently (1994/1995) high voltage p-channel Power MOSFETS were introduced commercially. One such device is IXTH11P50 from IXYS Semiconductors with a voltage rating of 500V and current rating of 11A and an on-resistance of 900mΩ.

Such high-current devices eliminate the need to parallel many lower-current FETs. These devices make possible complementary high-voltage push-pull circuits and simplified half-bridge and H bridge motor drives.

Recently introduced low voltage p MOSFETs from the TrenchFET family of Temic Semiconductors (Goodenough, 1997) have typical $r_{DS(on)}$ values between 14mΩ to 25mΩ.

2.5.8.3 More Advanced Power MOSFETS

With the advancement of processing capabilities, industry benefits with more advanced Power MOSFETS such as:

(a) Current sensing MOSFETS
(b) Logic-level MOSFETS

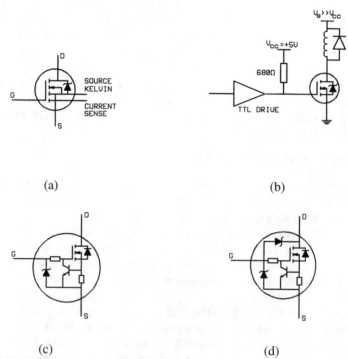

FIGURE 2-33 Advanced Monolithic MOSFETS (a) Current sensing MOSFET (b) Application of a logic-level MOSFET (c) Current limiting MOSFET (d) Voltage clamping, current limiting MOSFET

(c) Current limiting MOSFETS
(d) Voltage clamping, current limiting MOSFETS

The technique of current mirroring for source current-sensing purposes involves connecting a small fraction of the cells in a power MOSFET to a separate sense terminal. The current in this terminal (see Figure 2–33(a)) is a fixed fraction of the source current feeding the load. Current sense lead provides an accurate fraction of the drain current that can be used as a feedback signal for control and/or protection.

It's also valuable if you must squeeze the maximum switching speed from a MOSFET. For example, you can use the sense terminal to eliminate the effects of source-lead inductance in high-speed switching applications. Several manufacturers such as Harris, IXYS, Phillips, etc. manufacture these components.

Another subdivision of the rapidly diversifying power-MOSFET market is a class of devices called logic-level FETs. Before the advent of these units, drive circuitry had to supply gate-source turn-on levels of 10V or more. The logic-level MOSFETS accept drive signals from CMOS or TTL ICs that operate from a 5V supply. Suppliers of these types include International Rectifier, Harris, IXYS, Phillips-Amperex, and Motorola. Similarly other types described under (b) and (d) above are also available in monolithic form and some of there devices are categorized under "Intelligent Discretes."

2.6 Insulated Gate Bipolar Transistor (IGBT)

MOSFETs have become increasingly important in discrete power device applications due primarily to their high input impedance, rapid switching times, and low on-resistance. However, the on-resistance of such devices increases with increasing drain-source voltage capability, thereby limiting the practical value of power MOSFETs to application below a few hundred volts.

To make use of the advantages of power MOSFETs and BJTs together a newer device, Insulated Gate Bipolar Transistor (IGBT), has been introduced recently. With the voltage-controlled gate and high-speed switching of a MOSFET and the low saturation voltage of a bipolar transistor, the IGBT is better than either device in many high power applications. It is a composite of a transistor with an N-Channel MOSFET connected to the base of the PNP transistor.

Figure 2–34(a) shows the symbol and Figure 2–34(b) shows the equivalent circuit. Typical IGBT characteristics are shown in Figure 2–34(c). Physical operation of the IGBT is closer to that of a bipolar transistor than to that of a Power MOSFET. The IGBT consists of a PNP transistor driven by an N-channel MOSFET in a pseudo-Darlington Configuration.

The JFET supports most of the voltage and allows the MOSFET to be a low voltage type, and consequently have a low $R_{DS(on)}$ value. The absence of the integral reverse diode gives the user the flexibility of choosing an external fast recovery diode to match a specific requirement. This feature can be an advantage or a disadvantage, depending on the frequency of operation, cost of diodes, current requirement, etc.

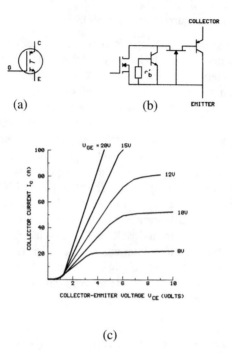

FIGURE 2-34 IGBT (a) Symbol (b) Equivalent circuit (c) Typical output characteristics

In IGBTs on-resistance values have been reduced by a factor of about 10 compared with those of conventional n-channel power MOSFETs of comparable size and voltage rating.

IGBT power modules are rapidly gaining applications in systems such as inverters, UPS systems and automotive environments. The device ratings are reaching beyond 1800 volts and 600A. The frequency limits from early values of 5KHz are now reaching beyond 20KHz while intelligent IGBT modules that include diagnostic and control logic along with gate drive circuits are gradually entering the market.

Characteristics comparison of IGBTs, Power MOSFETs, Bipolars, and Darlingtons are indicated in Table 2–2. References listed between Russel (1992) through Clemente, Dubhashi, and Pelly (1990) provide more details on IGBTs and their applications.

TABLE 2-2 Characteristics Comparison of IGBTs, Power MOSFETs, Bipolars, and Darlingtons

	Power MOSFETs	IGBTs	Bipolars	Darlingtons
Type of Drive	Voltage	Voltage	Current	Current
Drive Power	Minimal	Minimal	Large	Medium
Drive Complexity	Simple	Simple	High (large positive and negative currents are required)	Medium
Current Density for Given Voltage Drop	High at Low Voltage—Low at High Voltages	Very High (small trade-off with switching speed)	Medium (severe trade-off with switching speed)	Low
Switching Losses	Very Low	Low to Medium (depending on trade-off with conduction losses)	Medium to High (depending on trade-off with conduction losses)	High

2.7 MOS Controlled Thyristor (MCT)

MOS Controlled Thyristors are a new class of power semiconductor devices that combine thyristor current and voltage capability with MOS gated turn-on and turn-off. Various sub-classes of MCTs can be made: P-type or N-type, symmetric or asymmetric blocking, one or two-sided Off-FET gate control, and various turn-on alternatives including direct turn-on with light.

All of these sub-classes have one thing in common; turn-off is accomplished by turning on a highly interdigitated Off-FET to short out one or both of the thyristor's emitter-base junctions. The device, first announced a few years ago by General Electric's power semiconductor operation (now part of Harris Semiconductor, USA), was developed by Vic Temple. Harris is the only present (1997) supplier of MCTs, however ABB has introduced a new device called "Insulated Gate Commutated Thyristor" (IGCT) which is in the same family of devices.

Figure 2-35 depicts the MCT equivalent circuit. Most of the characteristics of an MCT can be understood easily by reference to the equivalent circuit shown here. MCT closely approximates a bipolar thyristor (the two transistor model is shown) with two opposite polarity MOSFET transistors connected between its anode and the proper layers to turn it on and off. Since MCT is a NPNP device rather than a PNPN device and output terminal or cathode must be negatively biased.

Driving the gate terminal negative with respect to the common terminal or anode turns the P channel FET on, firing the bipolar SCR. Driving the gate terminal positive with respect to the anode turns on the N channel FET on shunting the base drive to PNP bipolar transistor making up part of the SCR, causing the SCR to turn off. It is obvious from the equivalent circuit that when no gate to anode voltage is applied to the gate terminal of the device, the input terminals of the bipolar SCR are unterminated. Operation without gate bias is not recommended.

In the P-MCT a P-channel On-FET is turned on with a negative voltage which charges up the base of the lower transistor to latch on the MCT. The MCT turns on simultaneously over the entire device area giving the MCT excellent di/dt capability. Figure 2-36 compares different 600V power switching devices. Figure 2-37 compares the characteristics of 1000 V P-MCT device with a N-IGBT device of same voltage rating. Note that the MCT typically has 10 to 15 times the current capability at the same voltage drop.

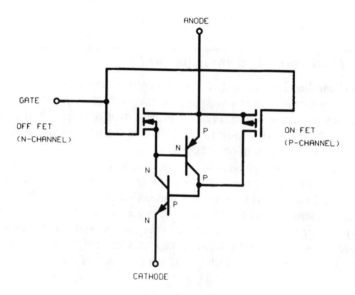

FIGURE 2-35 MCT equivalent circuit

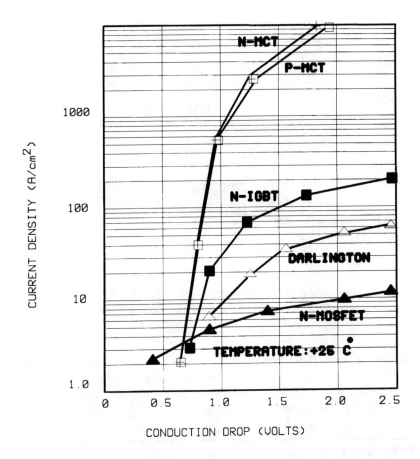

FIGURE 2-36 Comparison of 600 V devices (Copyright by Harris Corporation, reprinted with permission of Harris Semiconductor Sector)

The MCT will remain in the on-state until current is reversed (like a normal thyristor) or until the off-FET is activated by a positive gate voltage. Just as the IGBT looks like a MOSFET driving a BJT, the MCT looks like a MOSFET driving a thyristor (an SCR). SCRs and other thyristors turn on easily, but their turn-off requires stopping, or diverting virtually all of the current flowing through them for a short period of time. On the other hand, the MCT is turned off with voltage control on the high-impedance gate. The MCT offers a lower specific on-resistance at high voltage than any other gate-driven technology.

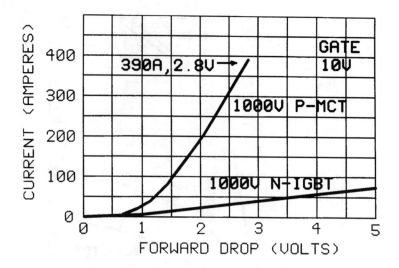

FIGURE 2-37 Comparison of forward voltage drop of 1000V P-MCT and N-IGBT at 150°C (Copyright by Harris Corporation, reprinted with permission of Harris Semiconductor Sector)

That is, just as the IGBT operates at a higher current density than the DMOSFET, the MCT (like all thyristors) operates at even higher current densities. In the future, the ultimate power switch may well be the MOS-controlled thyristor (MCT). References 16 to 20 provide details for designers.

References

1. Ashkianazi, G., J. Lorch, and M. Nathan. "Ultrafast GaAs Power Diodes Provide Dynamic Characteristics with Better Temperature Stability than Silicon Diodes." *PCIM*, April 1995, pp 10–16.
2. Hammerton, C. J. "Peak Current Capability of Thyristors." *PCIM*, November 1989, pp 52–55.
3. Coulbeck, L., W. J. Findlay, and A. D. Millington. "Electrical Trade-offs for GTO Thyristors." *Power Engineering Journal*, Feb. 1994, pp 18–26.
4. Bassett, Roger J. and Colin Smith. "A GTO Tutorial: Part I." *PCIM*, July 1989, pp 35–39.
5. Bassett, Roger J. and Colin Smith. "A GTO Tutorial: Part II - Gate Drive." *PCIM*, August 1989, pp 21–28.
6. McNulty, Tom. "Understanding Power MOSFETS." Harris Semiconductor, Application note AN 7244.2, Sept. 1993.
7. Travis, Bill. "Power MOSFETS & IGBTS." *EDN*, 5 January 1989, pp 128–142.
8. Goodenough, Frank. "Trench – Gate DMOSFETS In S0 – 8 Switch 10A at 30V." *Electronic Design*, 6 March 1995, pp 65–72.
9. Goodenough, Frank. "DMOSFETS Switch Milliwatts to Megawatts." *Electronic Design*, 5 September 1994, pp 57–65.
10. Furuhata, Sooichi and Tadashi Miyasaka. "IGBT Power Modules Challenge Bipolars, MOSFETS in Invertor Applications." *PCIM*, January 1990, pp 24–28.
11. Russel, J. P. et al. "The IGBTs—A new high conductance Mos-Gated device." Harris Semiconductor, App. note AN 8602.1, May 1992.

12. Wojslawowicz, J. E. "Third Generation IGBTS Approach Ideal Switch Capability." *PCIM*, January 1995, pp 28–32.
13. Frank, Randy and John Wertz. "IGBTS Integrate Protection for Distributorless Ignition Systems." *PCIM*, February 1994, pp 42–49.
14. Dierberger, Ken. "IGBT Do's and Don'ts." *PCIM*, August 1992, pp 50–55.
15. Clemente, S., A. Dubhashi, and B. Pelly. "Improved IGBT Process Eliminates Latch–up, Yields Higher Switching Speed – Part I." *PCIM*, October 1990, pp 8–16.
16. Temple, V., D. Watrous, S. Arthur, and P. Kendle. "Mos-Controlled Thyristor (MCT) Power Switches – Part I – MCT Basics." *PCIM*, November 1992, pp 9–16.
17. Temple, V., Watrous, D., S. Arthur, and P. Kendle. "Mos-Controlled Thyristor (MCT) Power Switches – Part II: Gate Drive and Applications." *PCIM*, January 1993, pp 24–33.
18. Temple, V., D. Watrous, S. Arthur, and P. Kendle. "Mos-Controlled Thyristor (MCT) Power Switches – Part III: Switching, Applications and The Future." *PCIM*, February 1993, pp 24–33.
19. Temple, V. A. K. "Mos Controlled Thyristors – A New Class of Power Devices." IEEE trans., *Electron Devices* (Vol. ED–33, No. 10), pp 1609–1618.
20. Temple, V. A. K. "Advances in MOS – Controlled Thyristor Technology." *PCIM*, November 1989, pp 12–15.
21. Burkel, R. and T. Schneider. "Fast Recovery Epitaxial Diodes Characteristics – Applications – Examples." IXYS Technical Information 33 (Publication No. D94004E, 1994).
22. Williams, B.W. *Power Electronics: Devices, Drivers, Applications and Passive Components.* Macmillan (1992).
23. Kinzer, Dan. "Fifth-Generation MOSFETs Set New Benchmarks for Low On- Resistance." *PCIM*, August 1995, pp 59.
24. Delaney, S., A. Salih, and C. Lee. "GaAs Diodes Improve Efficiency of 500kHz Dc-Dc converter." *PCIM*, August 1995, pp 10–11.
25. Deuty, S. "GaAs Rectifiers Offer High Efficiency in a 1MHz. 400Vdc to 48 Vdc Converter." HFPC Conference Proceedings, September 1996, pp 24–35.
26. Davis, C. "Integrated Power MOSFET and Schottky Diode Improves Power Supply Designs." *PCIM*, January, 1997, pp 10–14.
27. Goodenough, Frank. "Dense MOSFET enables portable power control." *Electronic Design*, April 14, 1997, pp 45 – 50.

Bibliography

1. Bradley, D. A. *Power Electronics.* Chapman & Hall, 1995.
2. Bird, B. M., K.G. King, and D.A.G. Pedder. *An Introduction to Power Electronics* (2nd Edition), John Wiley (1993).
3. Bose, B. K. *Modern Power Electronics.* IEEE Press, 1992.
4. Roehr, Bill. "Power Semiconductor Mounting Considerations." *PCIM*, September 1989, pp 8–18.
5. Polner, Alex. "Characteristics of Ultra High Power Transistors." Proceeding of first national solid state power conversion conference, March 1975.
6. Rippel, Wally E. "MCT/FET Composite Switch. Big Performance with Small Silicon." *PCIM*, November 1989, pp 16–22.
7. Frank, Randy and Richard Valentine. "Power FETS Cope with the Automotive Environment." *PCIM*, February 1990, pp 33–39.
8. Anderson, S., K. Gauen, and C.W. Roman. "Low Loss, Low Noise Diodes Improve High Frequency Power Supplies." *PCIM*, February 1991, pp 6–13.
9. Driscoll, J. "Bipolar Transistors and High Side Switches in High Voltage, High Frequency Power Supplies." Proceedings of power conversion conference, October 1990.
10. Schultz, Warren. "Ultrafast – Recovery Diodes Extend the SOA of Bipolar Transistors." *Electronic Design*, 14 March 1985, pp 167–174.
11. Sasada, Yorimichi, Shigeki Morita, and Makato Hideshima: "High Voltage, High Speed IGBT Transistor Modules." *Toshiba Review*, No 157 (Autumn) 1986, pp 34–38.

12. Peter, Jean Marie. "State of The Art and Development in the Field of Medium Power Devices." *PCIM*, May 1986, pp 14–22.
13. Smith, Colin and Roger Bassett. "GTO Tutorial Part III - Power Loss in Switching Applications." *PCIM*, September 1989, pp 99–105.
14. Mitlehner, H. and H. J. Schulze. "Current Developments in High Power Thyristors." *EPE Journal*, March 1994 (Vol 4, No.1), pp 36–42.
15. Lynch, Fernando. "Two Terminal Power Semiconductor Technology Breaks Current/ Voltage/Power Barrier." *PCIM*, October 1994, pp 10–14.
16. Deuty, Scott, Emory Carter, and Ali Salih. "GaAs Diodes Improve Power Factor Correction Boost Converter Performance." *PCIM*, January 1995, pp 8–19.
17. Goodenough, Frank. "DMOSFETS, IGBTS Switch High Voltage." *Electronic Design*, 7 November 1994, pp 95–105.
18. Adler, Michael, et al. "The Evolution of Power Device Technology." *IEEE Transactions on Electronic Devices*, November 1984, Vol ED–31, No 11, pp 1570–1591.
19. Ramshaw, R. S. *Power Electronics Semiconductor Switches*. Chapman & Hall, 1993.
20. International Rectifier."Schottky Diode Designer's Manual." 1992.
21. Mohan, N., T. M. Undeland, and W. P. Robbins. *Power Electronics: Converter, Applications and Design*. John Wiley, 1989.
22. Borras, R. P. Aloisi, and D. Shumate. "Avalanche Capability of Today's Power Semiconductors." Proceedings of EPE–93 (Vol. 2), 1993, pp 167–171.
23. Arthur, S. D. and V. A. K. Temple. "Special 1400 Volt N – MCT Designed for Surge Applications." Proceedings of EPE 93, (Vol 2) (1993), pp 266–271.
24. Gauen, K. and W. Chavez. "High Cell Density MOSFETS: Low on Resistance Affords New Design Options." Proceedings of PCI, October 1993, pp 254–264.
25. Driscoll, J. C. "High current fast turn-on pulse generation using Power Tech PG–5xxx series of "Pulser" gate assisted turn-off thyristors (GATO's)." Power Tech App Note, 1990.
26. Serverns, Rudy and Jack Armijos. "MOSPOWER Applications Handbook." Siliconix Inc., 1984.
27. Nilsson, T. "The insulated gate bipolar transistor response in different short circuit situations." Proceedings of EPE-93, 1993, pp 328–331.
28. Consoli, A., et al. "On the selection of IGBT devices in soft switching applications." Proceedings of EPE–93, pp 337–343.
29. Eckel, H. G. and L. Sack. "Optimization of the turn-off performance of IGBT at overcurrent and short-circuit current." Proceedings of EPE-93, 1993, pp 317–321.
30. Heumann, K. and M. Quenum. "Second Breakdown and Latch-up Behavior of IGBTs." EPE-93, 1993, pp 301–305.
31. Mitter, C. S. "Introduction to IGBTs." *PCIM*, December, 1995, pp 32–39.
32. Travis, B. "MOSFETs and IGBTs Differ in Drive Methods and Protections Needs." EDN, March 1, 1996, pp 123–132.
33. Locher, R. E. "1600V BIMOSFET™ Transistors Expand High Voltage Applications." *PCIM*, August 1996, pp 8–21.
34. Barkhordarian, V. "Power MOSFET Basics." *PCIM*, June 1996, pp 28–39.

CHAPTER **3**

DC to DC Converters

Co-Author: Dileeka Dias

3.1 Introduction

The linear power supply is a mature technology, having been used since the dawn of electronics. However, due to its low efficiency, the use of bulky heat sinks, cooling fans, and isolation transformers, this type of power supply tends to be unfit for most of today's compact electronic systems.

The disadvantages of linear power supplies are greatly reduced by the regulated switching power supply. In this technology, the AC line voltage is directly rectified and filtered to produce a raw high-voltage DC. This, in turn, is fed into a switching element which operates at a high frequency of 20 kHz to 1 MHz, chopping the DC voltage into a square wave. This high-frequency square wave is fed into the power isolation transformer, stepped down to a predetermined value and then rectified and filtered to produce the required DC output. This DC to DC conversion process which is at the heart of switching power supplies is examined in this chapter.

Switch mode power supplies have existed since the 1960s. Until the late 1970s, the task of efficiently converting one DC voltage to another using switching techniques was done with discrete components. Early systems were running at very low frequencies, usually between a few kHz to about 20 KHz, and were justified only for use in high power applications, usually above 500 watts.

The first ICs for switching power conversion were introduced in the late 1970s. During the mid 80s, switching techniques were introduced for much lower power ratings, while the operating frequency extended up to about 100 kHz. Additional enhancements such as multiple output rails also became available. Today, most switchers operate well above 500 kHz, with new magnetics, resonant techniques and surface mount technology extending this to several MHz. However, a major concern in such high-frequency switching power supplies is the minimization of the RFI pollution generated.

The recent rapid advancement of microelectronics has created a necessity for the development of sophisticated, efficient, lightweight power supplies which have a high power-to-volume (W/in^3) ratio, with no compromise in performance. High frequency switching power supplies, being able to meet these demands, have become the prime power source in a majority of modern electronic systems.

The combination of high efficiency (i.e., no large heat sinks) and relatively small magnetics, results in compact, light weight power supplies, with power densities in excess of 100 W/in^3 (Maliniak 1995 and Travis 1996) versus 0.3 W/in^3 for linear regulators. Still further innovations in DC/DC conversion are spurred by the proliferation of portable battery operated electronic devices and progressively plummeting logic voltages.

This chapter begins with a description of the basic principles of DC to DC conversion, and continues through more advanced resonant techniques to the latest technological innovations such as synchronous conversion and switched capacitor techniques. State of the art components are described as illustrative examples.

3.2 DC to DC Conversion Fundamentals

This section presents a brief overview of the DC to DC converter fundamentals, and a simplified analysis for characterizing the input/output relationship. More detailed descriptions can be found in (Brown 1990) and (Brown 1994). Ideal components are assumed for simplicity of analysis and description.

3.2.1 Modes of Operation

There are two major operational modes of the DC to DC converter within switching power supplies. They are the forward mode and the flyback mode. Although they have only subtle differences between them with respect to component arrangement, their operation is significantly different, and each has advantages in certain areas of application.

3.2.1.1 Forward Mode Converters

Figure 3-1(a) shows a simple forward mode converter. This type of converter can be recognized by an L-C filter section directly after the power switch or after the output rectifier on the secondary of a transformer. The Power Switch is a power transistor or a power MOSFET operated between fully-conduction and cut-off modes.

The operation of the converter can be seen by breaking its operation into two periods:

- **Power Switch ON period:** When the power switch is on, the input voltage is presented to the input of the L-C section, and the inductor current ramps upwards linearly. During this period the inductor stores energy.

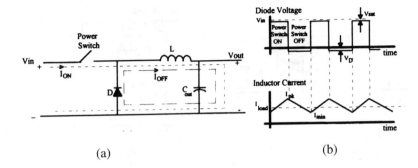

FIGURE 3-1 The forward mode converter (a) The basic circuit (b) Associated waveforms

- **Power Switch OFF period:** When the power switch is off, the voltage at the input of the inductor flies below ground since the inductor current cannot change instantly. Then the diode becomes forward biased. This continues to conduct the current that was formally flowing through the power switch. During this period, the energy that was stored in the inductor is dumped on to the load. The current waveform through the inductor during this period is a negative linear ramp. The voltage and current waveforms for this converter are shown in Figure 3-1(b).

The DC output load current value falls between the minimum and the maximum current values, and is controlled by the duty cycle. In typical applications, the peak inductor current is about 150 percent of the load current and the minimum is about 50 percent.

The advantages of forward mode converters are that they exhibit lower output peak-to-peak ripple voltages, and that they can provide high levels of output power, up to kilowatts.

3.2.1.2 Flyback Mode Converters

In this mode of operation, the inductor is placed between the input source and the power switch as shown in Figure 3-2. This circuit is also examined in two stages:

- **Power Switch ON period:** During this period, a current loop including the inductor, the power switch and the input source is formed. The inductor current is a positive ramp and energy is stored in the inductor's core.
- **Power Switch OFF period:** When the power switch turns off, the inductor's voltage flies back above the input voltage, resulting in forward biasing the diode. The inductor voltage is then clamped at the output voltage. This voltage which is higher than the input voltage is called the flyback voltage. The inductor current during this period is a negative ramp.

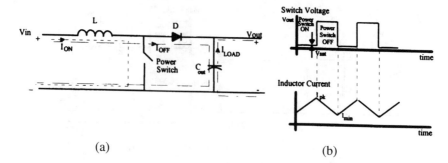

FIGURE 3-2 The flyback mode converter (a) The basic circuit (b) Associated waveforms

In Figure 3-2, the inductor current does not reach zero during the flyback period. This type of a flyback converter is said to operate in the continuous mode. The core's flux is not completely emptied during the flyback period, and a residual amount of energy remains in the core at the end of the cycle. Accordingly, there may be instability problems in this mode. Therefore, the discontinuous mode is the preferred mode of operation in flyback mode converters. The voltage and current waveforms are shown in Figure 3-3.

The only storage for the load in the flyback mode of operation is the output capacitor. This makes the output ripple voltage higher than in forward mode converters. The power output is lower than forward mode converters owing to higher peak currents that are generated when the inductor voltage flies back. As they consist of the fewest number of components, they are popular in low-to-medium power applications.

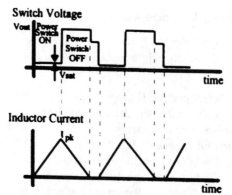

FIGURE 3-3 Voltage and current waveforms for the discontinuous flyback mode converter

3.2.2 Analysis of DC to DC Converters

3.2.2.1 The Volt-Second Balance for Inductors

The input-output characteristics of all DC/DC converters can be examined by using the requirement that the initial and the final inductor current within a cycle should be the same for steady-state operation (i.e., the fact that the net energy storage within one switching cycle in each inductor should be zero). This leads to the volt-sec balance for the inductor, which means that the average voltage per cycle across the inductor must be zero, i.e., that the volt-sec products for the inductor during each switching cycle should sum to zero (Brown 1990) and (Bose 1992).

This can be illustrated using a typical inductor current waveform as shown in Figure 3–4.

Let the positive slope of the current be m_1 and the negative slope m_2.
For the initial and final currents to be the same:

$$m_1 T_{on} - m_2 T_{off} = 0 \tag{3.1a}$$

Since the voltage across an inductor L is given by $L\, di/dt$,

$$V_{Lon} = Lm_1 \tag{3.1b}$$

$$V_{Loff} = -Lm_2 \tag{3.1c}$$

Combining (3.1a) – (3.1c)

$$V_{Lon} T_{on} / L + V_{Loff} T_{off} / L = 0 \tag{3.1d}$$

Equation (3.2) can be written as a function of the duty cycle D

$$D = T_{on} / (T_{on} + T_{off}) \tag{3.1e}$$

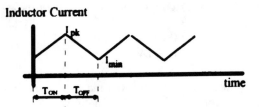

FIGURE 3–4 A typical inductor current waveform

Combining (3.1d) and (3.1e),

$$V_{Lon}D + V_{Loff}(1-D) = 0 \qquad (3.1f)$$

where V_{LON} and V_{LOFF} are the voltages across the inductor during the switch ON and the OFF periods respectively and D is the duty cycle.

The output voltage of forward mode converters is given approximately by $V_{out} = DV_{in}$, where D is the duty cycle of the switching waveform. Thus, this mode of operation always performs a step-down operation.

The output voltage for the flyback mode of operation is given by $V_{out} = V_{in}/(1-D)$. Thus, flyback mode converters are always used as step-up converters.

3.3 Converter Topologies

The converter topology refers to the arrangement of components within the converter. Converter topologies fall into two main categories: transformer-isolated and non-transformer-isolated. In each category, there are several topologies, with some available in both forms. An excellent discussion on converter topologies is found in (Bose 1992).

3.3.1 Non-Transformer-Isolated Converter Topologies

This type of switching converters are used when some external component such as a 50-60 Hz transformer or bulk power supply provides the DC isolation and protection. These are simple, thus easy to understand and design. However, they are more prone to failure due to lack of DC isolation. Hence these are used by designers mostly in situations such as distributed power systems, where a bulk power supply provides the necessary DC isolation.

There are four basic non-transformer-isolated topologies, the Buck (or step-down), the Boost (or step-up), the Buck-Boost (or inverting), and the Cuk converters.

The Buck Converter

This is the most elementary forward-mode converter, and is the building block for all forward-mode topologies. The buck converter topology is shown in Figure 3-5.

Applying equation (3.1f) to Figure 3-5, we see that for steady-state operation:

$$(V_{in} - V_{out})D - V_{out}(1-D) = 0$$

Hence, $$V_{out} = DV_{in}$$

As $V_{out} < V_{in}$ this is called the Buck or the step-down converter.

Though the buck converter is capable of delivering over 1 kW in normal operation, it is not a popular choice among experienced designers due to the following limitations:

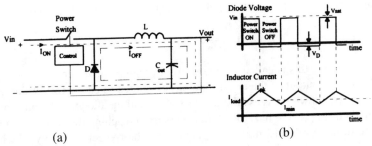

FIGURE 3-5 The buck converter topology (a) The basic circuit (b) Associated waveforms

i. The input voltage must always be 1 to 2 V higher than the output in order to maintain regulation. This is similar to the headroom voltage requirement in linear regulators. The headroom voltage is the actual voltage drop between the input voltage and the output voltage during operation. Ninety-five percent of all the power lost within the linear regulator is lost across this voltage drop.

ii. The finite reverse recovery time of the diode presents an instantaneous short circuit across the input source when the power switch turns on. This adds stress to the power switch and the diode.

iii. If the semiconductor power switch fails in the short-circuited condition (which is a common mode of failure for such devices when they do fail), the input is short-circuited to the load. Additional means of protection must be provided to prevent this as discussed in (Brown 1990).

The Boost Converter

The boost converter topology is shown in Figure 3-6. This is an elementary flyback-mode converter as can be seen by comparing Figure 3-2 with Figure 3-6.

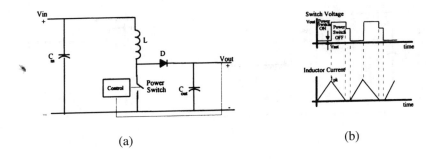

FIGURE 3-6 The boost converter topology (a) The basic circuit (b) Associated waveforms

This circuit can also be analyzed using equation (3.1f) which gives:

$$V_{in}D + (V_{in} - V_{out})(1-D) = 0$$

Hence,
$$V_{out} = V_{in}/(1-D) \qquad (3.3)$$

As $V_{out} > V_{in}$ this is called the boost or the step-up converter.

The power output of the boost topology is limited to about 150 W due to the high peak currents which stress the power switch and the diode. The ability of the boost regulator to prevent hazardous transients from reaching the load is also quite poor. Due to these reasons, many designers favor the flyback topology which is a transformer-isolated version of this converter. The flyback topology is discussed in Section 3.3.2.1.

The Buck-Boost Converter

The buck-boost converter topology is shown in Figure 3-7. This is a flyback-mode converter whose operation is very similar to that of the Boost topology. The difference between the two is that the positions of the inductor and the power switch have been reversed.

The inductor stores energy during the power switch's ON time and releases energy below ground through the diode into the output capacitor. This results in a negative voltage whose level is regulated by the duty cycle of the power switch. Analyzing this circuit using equation (3.1f) gives,

$$V_{in}D + V_{out}(1-D) = 0$$

Hence,
$$V_{out} = -V_{in}D/(1-D) \qquad (3.4)$$

From equation (3.4) it can be observed that, $|V_{out}| < V_{in}$ if $D < 50\%$ and $|V_{out}| > V_{in}$ if $D > 50\%$. Hence this is called the buck-boost converter. Since the output voltage is opposite in polarity to the input, it is also referred to as the inverting topology. However, due to the practical constraint of emptying the core's storage energy during the switch's OFF period, the duty cycle is limited to below 50% in most cases. The buck-boost topology also suffers from failure modes similar to those of the buck and the boost topologies (Brown 1990).

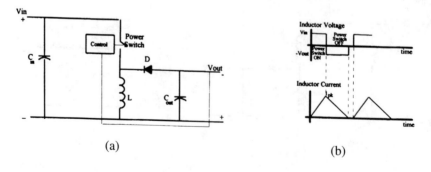

FIGURE 3-7 The buck-boost converter topology (a) The basic circuit (b) Associated waveforms

The Cuk Converter

The Cuk converter reduces the switching component of the input and the output currents, and hence the ripple to a great extent. This is done by having an inductor in series with both the input and the output for either switch position. The Cuk converter is shown in Figure 3-8.

When the power switch is OFF, the diode is ON and the capacitor charges with the current $I_{1\ OFF}$. The current $I_{2\ OFF}$ flows through the load and the diode.

When the power switch turns ON, the diode becomes reverse biased due to the capacitor voltage. The energy stored in the capacitor is dumped into the load through $I_{2\ ON}$. As the load current is allowed to flow continuously during either clock cycle, by suitable coupling of input and output inductors, the ripple in the input and output currents is very small. This can be nulled to zero by coupled inductor techniques as shown in (Chryssis 1989). The Cuk converter is based on capacitive energy transfer, unlike the other topologies, where inductors are used for energy transfer. The analysis of this topology is done by applying equation (3.1) to the two inductors in the circuit separately.

For inductor L_1 :

(3.5a)
$$V_{in}D + (V_{in} - V_c)(1-D) = 0$$
$$V_c = V_{in}/(1-D)$$

For inductor L_2 :

(3.5b)
$$(V_{out} + V_c)D + V_{out}(1-D) = 0$$
$$V_{out} = -DV_C$$

Using (3.5 a) and (3.5 b),

$$V_{out} = -V_{in}D/(1-D) \tag{3.5c}$$

Therefore, the output of the Cuk converter is buck-boost with polarity reversal.

The Cuk converter is closer to the ideal power supply compared to the other topologies. Therefore it has been referred to as an optimum switching power supply configuration, as it incorporates the advantages of switching technology and substantially eliminates the technology's main disadvantage, the switching currents. Also, the Cuk converter can be implemented in transformer-isolated forms as well, as discussed in Section 3.2.2.6. A detailed description and analysis of the Cuk converter is found in (Chryssis 1989) and (Cuk and Middlebrok 1994).

3.3.2 Transformer-Isolated Converter Topologies

In non-transformer-isolated converter topologies, only semiconductors provide DC isolation from input to output. The transformer-isolated switching converter category relies on a physical barrier to provide galvanic isolation. Not only can this withstand very high voltages before failure, but it also provides a second form of protection in the event of a semiconductor failure. Another advantage is the ease of adding multiple outputs to the power supply without separate regulators for each. These features make transformer isolated topologies, the preferred ones by designers.

In the transformer-isolated category, there are both forward and flyback mode converter topologies. The isolation transformer now provides a step-up or step-

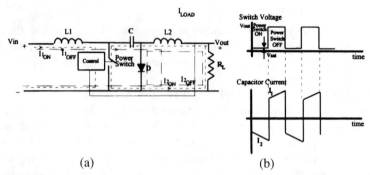

FIGURE 3-8 The Cuk converter topology (a) The basic circuit (b) Associated waveforms

down function. The topologies are called isolated forward, flyback, push-pull, half-bridge and full-bridge.

3.3.2.1 The Flyback Converter

The flyback converter topology and its waveforms are shown in Figure 3–9. This is the only flyback-mode converter in the transformer-isolated family of regulators. The flyback topology closely resembles the boost topology except for the addition of a secondary winding on the inductor. This flyback transformer is designed to perform the dual function of transformer and inductor.

During the power switch's on time, the full input is applied across the primary winding of the transformer, and the current flowing through it increases linearly. When the power switch turns off, the output diode starts to conduct, passing the energy stored within the core material to the capacitor and the load. During this flyback period, the secondary current is a negative ramp. The flyback topology can operate in either the discontinuous mode or the continuous mode.

Because of the flyback converter's unipolar use of the transformer core's B-H characteristics, the transformer core could get saturated. In such a condition, the power switch could fail due to the lowering of the primary inductance, and the

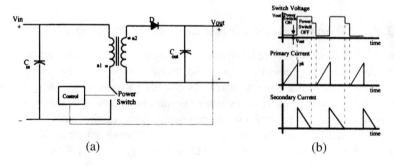

FIGURE 3-9 The Flyback Converter Topology (a) The basic circuit (b) Associated waveforms

resulting high primary current. The core should be designed to avoid this condition as described in (Brown 1990).

3.3.2.2 The Push-Pull Converter

The push-pull is a double ended topology (i.e., two power switches share the switching function). This forward-mode converter is shown in Figure 3-10. It has similarities to the buck converter.

The input is applied to the center-tap of the primary of the transformer, and two power switches are connected across each section of the primary winding. The secondary voltage is full-wave rectified and applied to a buck-style LC section.

The two switches share the switching function by alternately turning on and off. When each switch turns on, current flows through its side of the primary. Simultaneously, one half of the center-tapped secondary begins to conduct, forward biasing its respective diode, and letting the current into the load. Unlike in the flyback topology, the transformer does not store energy, but output current is drawn when either switch is on.

As the current flows in opposite directions in the primary winding during each switch's on period, the core material is driven in both the positive and the negative flux polarities. This reduces the possibility of core saturation, and also allows reduction of the core size by using it more efficiently. Another advantage is that this topology can provide twice the output power of a topology using one power switch operating at the same frequency. This feature enables the push-pull topology to generate output powers of several hundred watts.

For correct operation of this converter, there must be a dead time where neither transistor is conducting, to allow for the finite time necessary for the power switches to turn off (around 2 µS for bipolar transistors and 50–400 nS for MOSFETs). If both the switches are allowed to conduct at the same time, a short circuit will be formed in the transformer, resulting in high levels of current which could destroy the power switches.

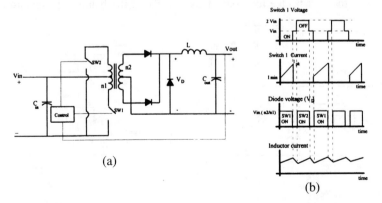

FIGURE 3-10 The push-pull converter topology (a) The basic circuit (b) Associated waveforms

66 POWER ELECTRONICS DESIGN HANDBOOK

This topology however, suffers from one serious flaw, as neither the two halves of the winding nor the two power switches can be absolutely identical due to real-world factors. Any asymmetry will eventually lead to saturation of one side of the transformer and to the destruction of the corresponding power switch. Some methods of overcoming this core imbalance are discussed in (Brown 1990). The topologies discussed in the remainder of this section eliminate this problem.

3.3.2.3 The Half-Bridge Converter

The half-bridge topology is also double ended. However, the half-bridge topology has only one primary winding which is connected between a pull-up/pull-down configuration of power switches and the centre node between two series capacitors as shown in Figure 3-11.

The capacitor centre node voltage is fixed approximately at half the input voltage. The other end of the primary is alternately presented with the input voltage and ground by the two power switches. Therefore, only half the input voltage appears across the primary winding at any time.

The switches alternate as before, and the direction of current alternates with each conducting switch. Hence the transformer core is operated in a bipolar fashion.

This topology exhibits a self-core balancing ability, which is accomplished by the capacitors. The centre node voltage will adjust itself according to the direction of higher flux density within the transformer. This reduces the voltage across the primary in the direction of impending saturation.

Since only half the input voltage appears across the primary, the peak current in the half-bridge topology is twice as high as in the push-pull topology with identical power output. Thus, this topology is not suited for as high power applications as the push-pull topology.

3.3.2.4 The Full-Bridge Converter

This is also a double ended topology. Its power output is significantly higher than that of the half-bridge topology. This is because the balancing capacitors are

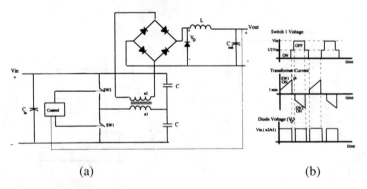

FIGURE 3-11 The half-bridge converter topology (a) The basic circuit (b) Associated waveforms

replaced with another pair of switches, identical to the first pair. This converter is shown in Figure 3-12.

Two of the four power switches are turned on simultaneously during each conduction cycle (e.g., SW1 and SW4 or SW2 and SW3). This places the full input voltage across the primary winding, which reduces the peak currents flowing through it compared to the half-bridge topology.

This effectively doubles the power handling capacity of this topology over the half-bridge. Therefore this is used in applications requiring output powers of 300 W to several kilowatts.

Core balancing may be achieved in this topology by placing a small, nonpolarized capacitor in series with the primary winding (Brown 1990). The average DC voltage across this capacitor causes to bring the core away from saturation. Core balancing is essential in this topology, as core saturation would lead to instant destruction of semiconductors at the power levels at which they operate.

3.3.2.5 The Transformer-Isolated Cuk Converter

The transformer-isolated version of the Cuk converter is shown in Figure 3-13. Comparing this with the basic Cuk converter in Figure 3–8 it is seen that DC isolation is achieved at the expense of one additional capacitor besides the isolation transformer.

The same charge balance of the non-isolated version is valid for capacitors C_A and C_B. Thus, average currents in both transformer windings are zero. This is unlike the flyback topology shown in Figure 3-9 where a large DC bias current is carried by the transformers. Therefore the transformer in the Cuk converter has advantages in terms of compactness and leakage reactances. As this circuit retains all the properties of the basic Cuk converter, the zero ripple extension using coupled inductor technique can be applied to this also as described in (Chryssis 1989).

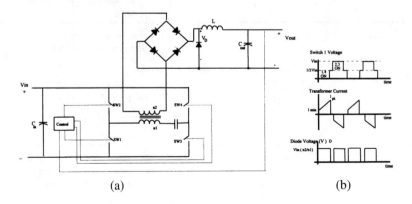

FIGURE 3-12 The full-bridge converter topology (a) The basic circuit (b) Associated waveforms

68 POWER ELECTRONICS DESIGN HANDBOOK

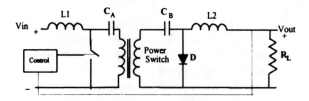

FIGURE 3-13 The transformer-isolated cuk converter topology

TABLE 3-1 Comparison of Converter Topologies

Topology	Vin /Vout	Advantages
Buck	D	High efficiency, simple, low switch stress, low ripple
Boost	$1/(1-D)$	High efficiency, simple, low input ripple current
Buck-boost	$-D/(1-D)$	Voltage inversion without a transformer, simple, high frequency operation
Flyback	$N_2D/N_1(1-D)$	Isolation, low parts count, has no secondary output inductors
Push-pull	$2N_2D/N_1$	Isolation, good transformer utilization, good at low input voltages, low output ripple
Half-bridge	N_2D/N_1	Isolation, good transformer utilization, switches rated at the input voltage, low output ripple
Full-bridge	$2N_2D/N_1$	Isolation, good transformer utilization, switches rated at the input voltage, low output ripple

3.3.3 Summary and Comparison of Converter Topologies

The converter topology has a major bearing on the conditions in which the power supply can operate safely, and on the amount of power it can deliver. Cost vs. performance trade-offs are also needed in selecting a suitable converter topology for an application.

The primary factors that determine a choice of topology are whether DC isolation is needed, the peak currents and voltages that the power switches are subjected to, the voltages applied to transformer primaries, the cost and the reliability.

Some relative merits of the converter topologies as well as their advantages, disadvantages and typical applications are summarized in Table 3-1. Mathematical expressions for estimating some of the important parameters such as peak currents and voltages, output power etc. are found in (Brown 1990) and (Brown 1994). A comparison of the strengths and weaknesses of the different converter topologies are also discussed in (Moore 1994).

TABLE 3-1 Continued

Disadvantages	Typical Application Environment	$V_{in}(dc)$ Range (V)	Typical Efficiency (%)	Relative Parts Cost
No isolation, potential over-voltage if switch shorts	Small sized imbedded systems	5 - 1000	78	1.0
No isolation, high switch peak current, regulator loop hard to stabilize, high output ripple, unable to control short-circuit current	Power-factor correction, battery up-converters	5 - 600	80	1.0
No isolation, regulator loop hard to stabilize, high output ripple	Inverse output voltages	5 - 600	80	1.0
Poor transformer utilization, high output ripple, fast recovery diode required	Low output power, multiple output	5 - 500	80	1.2
Cross conduction of switches possible, high parts count, transformer design critical, high voltage required for switches	Low output voltage	50 - 1000	75	2.0
Poor transient response, high parts count, cross conduction of the switches possible	High input voltage, moderate to high power	50 - 1000	75	2.2
High parts count, cross conduction of switches possible	High power, high input voltage	50 - 1000	73	2.5

The industry has settled into several primary topologies for a majority of the applications. Due to reasons discussed in Section 3.3.2, transformer-isolated topologies have become more popular than non-transformer-isolated topologies.

Figure 3-14 illustrates the approximate range of usage for these topologies. The boundaries to these areas are determined primarily by the amount of stress the power switches must endure and still provide reliable performance. The boundaries delineated in Figure 3-14 represent approximately 20A of peak current.

The flyback configuration is used predominantly for low to medium output power (150W) applications because of its simplicity and low cost. Unfortunately, the flyback topology exhibits much higher peak currents than do the forward-mode supplies. Therefore, at higher output powers, it quickly becomes an unsuitable choice. For medium-power applications (100 to 400 W) the half-bridge topology becomes the predominant choice. The half-bridge is more complex than the flyback and therefore costs more.

However, its peak currents are about one-third to one-half of those exhibited by the flyback. Above 400 W the dominant topology is the full-bridge, which offers the most effective utilization of the full capacity of the input power source. It is also the most expensive to build, but for those power levels the additional cost becomes a trivial matter. Above 150 W the push-pull topology is also sometimes used. However, this exhibits some shortcomings such as core imbalance, that may make it tricky to use (Brown 1990).

Estimating the major power supply parameters at the beginning of the design and by using charts such as Figure 3-14 will lead to the selection of a final topology which is safe, reliable and cost-effective.

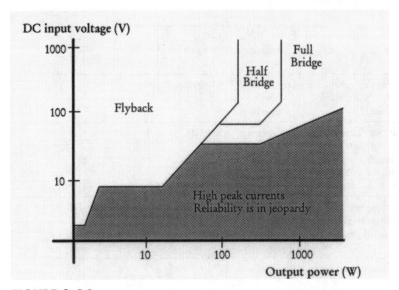

FIGURE 3-14 Industry favorite configurations and their areas of usage

3.4 Control of DC to DC Converters

In a PWM switch mode power supply, a square wave pulse is normally generated by the control circuit to drive the switching transistor ON and OFF. By varying the width of the pulse, the conduction time of the transistor is correspondingly varied, thus regulating the output voltage. The major function of the control subsection of a PWM supply is thus to sense any change in the DC output voltage and adjust the duty cycle of the power switches to correct for such changes.

An oscillator sets the basic frequency of operation of the power supply. A stable, temperature-compensated reference is used to which the output voltage is compared in a high-gain voltage error amplifier. An error-voltage to pulse width converter is used to adjust the duty cycle.

The PWM control circuit may be single ended for driving single-transistor converters such as the buck or boost topologies, or it may be double ended to drive multiple-transistor converters such as the push-pull or half-bridge topologies.

There are two basic modes of control used in PWM converters: voltage mode and current mode. Variable frequency control techniques used mostly in resonant power supplies are discussed in a subsequent section.

3.4.1 PWM Control Techniques

3.4.1.1 Voltage-Mode Control

This is the traditional mode of control in PWM switching converters. It is also called single-loop control as only the output is sensed and used in the control circuit. A simplified diagram of a voltage-mode control circuit is shown in Figure 3–15.

The main components of this circuit are an oscillator, an error amplifier, and a comparator. The output voltage is sensed and compared to a reference. The error voltage is amplified in a high-gain amplifier. This is followed by a comparator which compares the amplified error signal with a saw-tooth waveform generated across a timing capacitor.

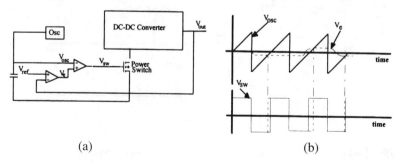

FIGURE 3–15 Voltage-mode control (a) Block diagram (b) Associated waveforms

The comparator output is a pulse-width modulated signal which serves to correct any drift in the output voltage. As the error signal increases in the positive direction, the duty cycle is decreased, and as the error signal increases in the negative direction, the duty cycle is increased.

The voltage mode control technique works well when the loads are constant. If the load or the input changes quickly, the delayed response of the output poses a drawback to the control circuit as it only senses the output voltage. Also, the control circuit cannot protect against instantaneous overcurrent conditions on the power switch. These drawbacks are overcome in current-mode control.

3.4.1.2 Current-Mode Control

This is a multiloop control technique, which has an AC current feedback loop in addition to the voltage feedback loop. This second loop directly controls the peak inductor current with the error signal rather than controlling the duty cycle of the switching waveform. Figure 3-16 shows a block diagram of a basic current-mode control circuit.

The error amplifier compares the output to a fixed reference. The resulting error signal is then compared with a feedback signal representing the switch current in the current sense comparator. This comparator output resets a flip flop which is set by the oscillator. Therefore, switch conduction is initiated by the oscillator and terminated when the peak inductor current reaches the threshold level established by the error amplifier output.

Thus the error signal controls the peak inductor current on a cycle-by-cycle basis. The level of the error voltage dictates the maximum level of peak switch cur-

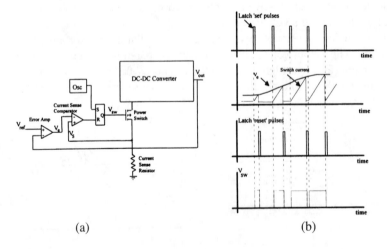

FIGURE 3-16   Current mode control (a) Block diagram (b) Associated waveforms

rent. If the load increases, the voltage error amplifier allows higher peak currents. The inductor current is sensed through a ground-referenced sense resistor in series with the switch.

The disadvantages of this mode of control are loop instability above 50% duty cycle, less than ideal loop response due to peak instead of average current sensing, tendency towards subharmonic oscillation and noise sensitivity, particularly at very small ripple current.

However, with careful design as explained in (Brown 1990), (Chryssis 1989), etc., these disadvantages can be overcome. Therefore, current-mode control becomes an attractive option for high frequency switching power supplies.

The comparator compares the feedback pin voltage with the reference signal. When the feedback voltage drops below the reference, the comparator switches on the oscillator. The driver amplifier boosts the signal level to drive the output power switch.

3.4.1.3 Gated Oscillator Control

Conventional PWM control techniques are fixed-frequency variable-pulse-width techniques. However, a new method of control for PWM converters is the gated oscillator method which turns the oscillator on with fixed pulse widths and turns it off using an overcurrent condition. The off-time is used to control the output. Figure 3–17 shows a block diagram of this control method.

The switch cycling action raises the output voltage and hence the feedback voltage. When the feedback voltage rises above the reference, the oscillator is gated off. The I_{lim} input sets the maximum switch current. When the maximum limit is exceeded, the switch is turned off.

The switch current is generally detected by sampling a small fraction of it and passing it through a resistor. Since the switch is cycled only when the feedback voltage drops below the reference voltage, this type of architecture has a very low supply current, and hence is suitable only for very low power applications (Brown 1994), (Linear Technology Corp. 1992).

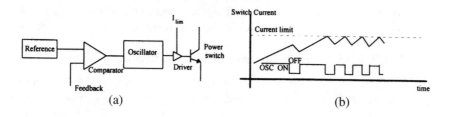

FIGURE 3–17 Gated oscillator control (a) Block diagram (b) Associated waveforms

3.4.2 Control Integrated Circuits

A wide variety of ICs have been developed to implement the control functions in switching power supplies. In addition to the basic functions of control circuits described above, these ICs provide more advanced functions and facilities such as:

(i) Overcurrent protection.
(ii) Soft-start facilities that start the power supply in a smooth fashion, reducing the inrush current.
(iii) Dead-time control which fixes the maximum pulse width, to prevent the simultaneous conduction of two power switches.
(iv) Undervoltage lockout to prevent the supply from starting when there is insufficient voltage within the control circuit for driving the power switches to saturation.

Selection of a control IC for a particular application depends on the power switch type, the intended topology, and the control mode as well as on the required output power, peak currents, and voltages. Another important factor is whether the power transistor is integrated on the control IC die itself.

Popular switching power supply control ICs, their features and applications are described in (Maliniak 1995), (Brown 1990), (Brown 1994), (Chryssis 1989), and (Linear Technology Corp. 1990).

3.4.2.1 A Typical Voltage-Mode Controller

Figure 3–18 shows the functional block diagram of the TL494 voltage mode controller from Motorola Inc. (Motorola, Inc., 1997). Control for both push-pull and single-ended operation can be achieved with this chip. The frequency of the oscillator is set by the external resistors R_T and C_T. The typical operating frequency is 40 kHz, and the maximum is 200 kHz.

Output pulse width modulation is accomplished by comparison of the positive sawtooth waveform across the capacitor to either of two control signals. The NOR gates, which drive the output transistors are enabled only when the flip-flop clock input is in its low state. This happens only during that portion of time when the sawtooth voltage is greater than the control signals. Therefore, an increase in control signal amplitude causes a corresponding linear decrease of output pulse width.

3.4.2.2 A Typical Current-Mode Controller

Figure 3–19 shows the functional block diagram of the MAX747 current mode controller from Maxim Integrated Circuits (Maxim Integrated Products 1994). This is a CMOS step-down controller, which drives external P-channel FETs. This IC operates in continuous mode under heavy loads, but in discontinuous mode at light loads.

DC to DC Converters **75**

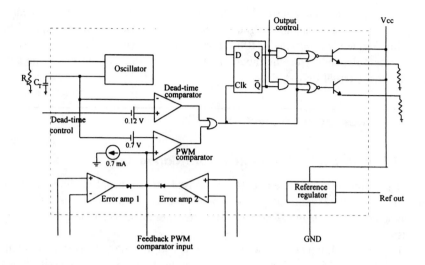

FIGURE 3-18 The functional block diagram of the TL 494 Voltage Mode Controller IC (Copyright of Motorola, used by permission)

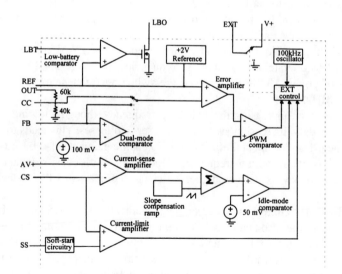

FIGURE 3-19 The functional block diagram of the MAX747 Current-Mode Controller IC (Reproduced by permission of Maxim Integrated Products)

Stability of the inner current-feedback loop is provided by a slope-compensation scheme that adds a ramp signal to the current-sense amplifier output. The switching frequency is nominally 100 kHz and the duty cycle varies from 5 percent to 96 percent depending on the input/output voltage ratio. EXT provides the gate drive for the external FET.

3.5 Resonant Converters

3.5.1 Introduction to Resonant Converters

The converters discussed so far, processed power in pulsed form, and are called Pulse Width Modulated (PWM) converters. The term resonance refers to a continuous sinusoidal signal. Resonant converters are those which process power in a sinusoidal form.

Resonant techniques have long been used with thyristor converters, in high-power SCR motor drives and in UPSs. However, due to its circuit complexity, it had not found application in low-power DC to DC converters until recently. However, due to certain advantages inherent in resonant techniques, and also due to the development of surface mount technology, resonant forms of DC to DC converters have gained rapid acceptance in this field of application.

The thrust towards resonant mode power supplies has been fueled by the industry's demand for miniaturization, together with increasing power densities and overall efficiency, and low EMI.

With available devices and circuit technologies, PWM converters have been designed to operate with switching frequencies the range 50–200 kHz. With the advent of power MOSFETs enabled the switching frequencies to be increased to several MHz.

However, increasing the switching frequency, though allowing for miniaturization, leads to increasing switching stresses and losses. This leads to reduction in efficiency. The detrimental effects of the parasitic elements also become more pronounced as the switching frequency is increased. With quasi-resonant techniques, higher frequency as well as higher efficiency compared to PWM techniques are achieved.

Resonant circuits in power supplies operate in two modes that define the flow of current in the resonant circuit: continuous and discontinuous. In the continuous mode, the circuit operates either above or below resonance. The controller shifts the frequency either towards or away from resonance, using the slope of the resonant circuit's impedance curve to vary the output voltage. This is a truly resonant technique, but is not commonly used in power supplies due to its high peak currents and voltages.

In the discontinuous mode, the control circuit generates pulses having a fixed on-time, but at a varying frequency determined by the load requirements. This mode of operation does not generate continuous current flow in the tuned circuit. This is the common mode of operation in a majority of resonant converters, and is called the quasi-resonant mode of operation.

3.5.2 The Quasi-Resonant Principle

The quasi-resonant principle is used in power converters by incorporating a resonant LC circuit with the power switch. The power switch is turned on and off in the same manner as in PWM converters, but the tank circuit forces the current through the switch into a sinusoidal form. The actual conduction period of the switch is governed by the resonant frequency f_r of the tank circuit. This basic principle is illustrated in Figure 3-20.

The tank circuit exhibits a relatively fixed ringing period to which the conduction period of the power switch is slaved. The ON period of the power switch is fixed to the resonance period of the tank elements. The quasi-resonant supply is controlled by changing the number of ON times of the power switch per second.

All resonant control circuits keep the pulse width constant and vary the frequency, whereas all PWM control circuits keep the frequency constant and vary the pulse width.

The main advantages of the quasi-resonance techniques arise from the sinusoidal waveshapes of the switching currents and voltages. The switching losses are reduced, leading to higher efficiency and EMI is greatly reduced.

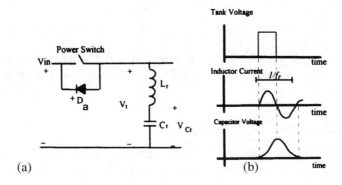

FIGURE 3-20 The resonant principle (a) The basic circuit (b) Associated waveforms

3.5.3 PWM vs. Quasi-Resonant Techniques

The operation of a quasi-resonant switching power supply is analogous to a PWM supply of the same topology. The difference lies in the fact that the switching waveform in quasi-resonant supplies has been preshaped into a sinusoidal form.

Figure 3–21 compares the switching waveforms of PWM and resonant converters. With PWM converters, there is simultaneous conduction of current and voltage during part of the switching period. In resonant conversion, switching can be achieved either at the zero current point or the zero voltage point of the sinusoidal switching waveform, thus minimizing the switching losses. However, increasing the switching frequency though allowing for miniaturization, is accompanied by increasing switching stresses, and detrimental effects due to parasitic elements.

The advantages and disadvantages of PWM and Quasi-Resonant conversion techniques are summarized in Table 3–2.

TABLE 3-2 Advantages and Disadvantages of Quasi Resonant Converters (QRCs) compared to PWM converters

Advantages	Disadvantages
• Lower switching losses, hence higher efficiency	• More complex circuit. Requires a longer design time and a higher level of expertise from the designer.
• Lower EMI	• Parasitic characteristics of components must be taken into account.
• Higher maximum operating frequencies.	• As current waveforms in quasi-resonant converters are sinusoidal, peak values are higher than those found in PWM converters.
• Hence smaller size of components	• Increased device stress.

The major applications in which quasi-resonant technology finds acceptance today are those that require very low EMI levels, such as in aircraft, satellite, and audio/video equipment.

3.5.4 The Resonant Switch

The resonant switch consists of a semiconductor switch and resonant LC elements. Because resonant circuits generate sinusoidal waves, designers can operate the power switches either at zero current or at zero voltage points in the resonant waveform. Based on this, there are two types of resonant switches: zero current switches (ZCS) and zero voltage switches (ZVS). The two types of switches are the duals of each other. A description of the operations of these switches is also found in (Brown 1990).

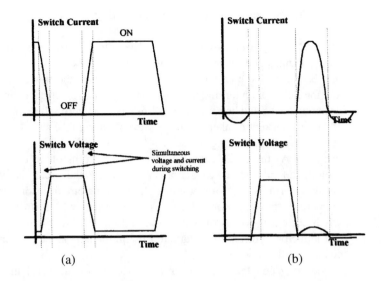

FIGURE 3-21 Comparison of switching waveforms in PWM and resonant converters (a) PWM switching waveforms (b) Resonant switching waveforms

3.5.4.1 Zero Current Quasi-Resonant Switches (ZC-QRCs)

A zero current quasi-resonant switch is shown in Figure 3-22. The switch type shown is called a full wave switch where the antiparallel diode D_a allows current to flow in either direction through the resonant circuit. Switches without such a diode allow only unidirectional current flow, and are called half wave switches. The operation of the switch can be broken down into four periods as shown in Figure 3-22.

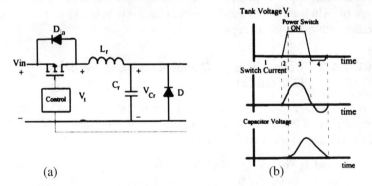

FIGURE 3-22 The zero-current quasi-resonant switch (a) The basic circuit (b) Associated waveforms

- **Period 1:** The power switch is off, and the diode D is conducting the load current.
- **Period 2:** The power switch turns on. The voltage across the switch makes a step change. The resonant capacitor appears to be short circuited at this time because of the conducting diode. Therefore, the power switch sees only the inductor on turning on. Therefore, the switch current cannot change instantaneously, and hence increases linearly from zero. This continues until all the load current is taken up by the current through the switch and the resonant inductor, displacing the current through the diode.
- **Period 3:** As the diode current is displaced, it turns off in a zero-current fashion, and the resonant capacitor is released into the circuit. Now, the current waveform assumes a sinusoidal shape as the circuit resonates. During this period, the capacitor voltage lags the current waveform by 90°. The switch current proceeds over its crest, and passes through zero. The resonant inductor's current then starts to flow in the opposite direction through the antiparallel diode D_a.
- **Period 4:** When the inductor current passes through zero, the resonant capacitor begins to dump its charge into the load, thus reducing its voltage in a linear ramp. The diode begins to conduct. When the capacitor voltage reaches zero, the diode takes up the entire current, and the circuit awaits the next conduction period of the power switch.

As the power switch and diode operate at zero current on both edges, switching losses in the power semiconductors are greatly reduced. The ON period of the power switch is the resonant period of the tank circuit. The number of ON times per second is varied by the control circuits. This method of control is called the fixed on-time, variable off-time method.

For heavier loads, the number of resonant ON periods is higher. Therefore, to increase the output power, the controller increases the frequency. Converters using this type of switch are called parallel resonant converters because the load is placed in parallel with the resonant capacitor.

3.5.4.2 Zero Voltage Quasi-Resonant Switches (ZV-QRCs)

A zero voltage quasi-resonant switch is shown in Figure 3–23. In this type, the power switch's OFF time is the resonant period. The operation of the switch can be broken down into four periods as shown.

- **Period 1:** The power switch is on. The switch current is determined by the converter stage configuration. The resonant inductor is saturated and is effectively short circuited. The input voltage appears across the resonant capacitor, and the diode D is off.
- **Period 2:** The resonant period is initiated by the power switch turning off. As the voltage across the capacitor cannot change instantaneously, the power switch voltage remains constant while the current reduces to zero.

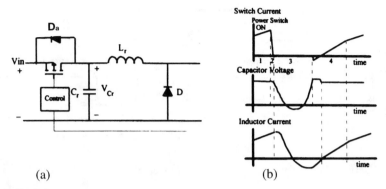

FIGURE 3-23 The zero-voltage quasi-resonant switch (a) The basic circuit (b) Associated waveforms

- **Period 3:** The capacitor voltage starts falling together with the inductor current. The diode starts to conduct, taking over the load current being from the resonant inductor which gradually falls out of saturation. The tank circuit begins to resonate. The capacitor voltage rings back above the input voltage, at which point the current is conducted by the antiparallel diode.
- **Period 4:** The power switch turns on. The diode D is also on at this time, and the capacitor is shunted out of the circuit. Therefore the switch current increases linearly through the resonant inductor. When this current exceeds the load current being conducted through the diode, the diode turns off. Then the resonant inductor can enter saturation and await the next cycle.

The number of OFF times per second is varied by the control circuits. This method of control is called the fixed off-time, variable on-time method. The "sense" of control is opposite to that in the ZCS. For heavier loads, the number of resonant OFF periods is decreased. Converters using this type of switch are called series resonant converters because the load is placed in series with the resonant inductor.

3.5.5 Quasi-Resonant Switching Converters

Comparable converter topologies as with the PWM technique are available with ZCS and ZVS quasi-resonant switching techniques. Circuits for each of these topologies are found in (Brown 1990), (Brown 1994), (Cuk and Middlebrook 1994), and (Steigerwald 1984). The available topologies for each type of switch are illustrated in Figure 3–24 A quasi-resonant version of the Cuk converter is described in (Cuk and Maksimovic 1988).

ZCS-QR supplies present a high current stress upon the power switches and ZVS-QR supplies a high voltage stress. This causes some topologies to be better suited for certain ranges of input voltages and output powers. ZCS-QR supplies are

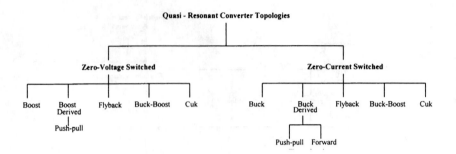

FIGURE 3-24 Quasi-resonant converter topologies

good at high input voltages, but poor at high output powers. Conversely, ZVS-QR supplies are good at high output powers but poor at high input voltages.

For off-line applications and for switching frequencies up to 1 MHz, the ZCS technique is effective as it eliminates the switching stresses and the turn-off stresses. However, the capacitive turn-on loss becomes high as the frequency is increased.

The ZVS technique minimizes the turn-on loss. However, the ZVS-QR switch suffers from two major limitations. One is due to the extensive voltage stress in single-ended configurations with wide load variations. Typically this stress is proportional to the load range, making them impractical for applications with a wide load range and/or high input current (Steigerwald 1984). The other limitation is related to the junction capacitance of the diode, which may cause undesirable oscillation with the resonant inductor.

Zero Voltage Switching QR converters appear to be the more popular of the two quasi-resonant techniques. One reason for this is that the typical variation in frequency over its input and load variation is 4:1 for ZVS as opposed to 10:1 for ZCS. Secondly, they have a better heavy load performance, and their parasitic elements are more easily controlled. Most ZCR applications handle no more than 300 W of output power, whereas ZVS applications handle powers in excess of many kilowatts.

3.5.6 Control Techniques and ICs for Resonant Converters

Control methods for resonant converters are variable frequency ones. Either the on-time is fixed and the off time is variable or vice-versa. The basic functioning of a resonant mode control circuit is shown in Figure 3–25. The fundamental blocks are a wide-band error amplifier, a voltage controlled oscillator (VCO) and a temperature-stable one-shot timer.

The output voltage is compared with the reference, and the error voltage is used to drive the VCO. The VCO output triggers the one-shot whose pulse duration is fixed as required by the converter. These control techniques are based on the voltage-mode of control, and hence suffer from poor input transient response characteristics.

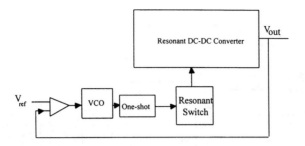

FIGURE 3-25 Resonant mode control

Some representative control ICs currently available are MC34066, LD405, UC1860, and UC3860. One of the first resonant mode controller ICs to appear on the market was the LD 405 by Gennum Corp. Subsequently, the CS3805 by Cherry Semiconductors and an improved GP605 by Gennum Corp. were released.

The LD405 and the CS3805 have drive current capabilities of 200 mA and operating frequencies in the range 10kHz to 1 MHz. They have single-ended and complementary outputs for driving power MOSFETs. The dissipation rate is specified at 500 mW at 50° C. The operating range of the GP605 is between 1 kHz and 1.2 MHz.

Unitrode Integrated Circuits introduced its UC3860 family of resonant mode controller chips in 1988. These were able to supply 800 mA, about 4 times the drive current and operate at higher frequencies (up to 2 MHz) than the previously available ICs. Their power dissipation rate is specified at 1.25W at 50° C.

A newer series of resonant mode control chips by Unitrode Integrated Circuits and their applications are described in (Wofford 1990). These are the UC3860, UC1861 (dual ZVS), UC1864 (single ZVS), and UC1865 (dual ZCS).

The functional block diagram of the UC1860 from Unitrode Integrated Circuits is shown in Figure 3-26 (Unitrode Integrated Circuits 1995-1996). The nominal operating frequency for this IC is 1.5 MHz, and implements resonant-mode fixed-on-time control, as well as a number of other power supply control schemes with its various dedicated and programmable features. This IC contains dual high-current totem-pole output drivers which can be programmed to operate alternately or in unison (Unitrode Integrated Circuits 1995-1996).

Although varying in detail and complexity all of these resonant mode control chips provide the same basic functions. All contain a VCO that varies the operating frequency, a monostable circuit that establishes the pulse on-time, and a steering circuit that determines the output drive mode (controlled on- or off-times, single-ended or complementary). The chips also provide many of the same basic protection features such as soft start, undervoltage lockout and overload protection—although in different ways and with varying degrees of sophistication.

Some of these control ICs are also described in (Brown 1990), (Brown 1994), and (Linear Technology Corp. 1994)-(Unitrode Integrated Circuits 1995-1996).

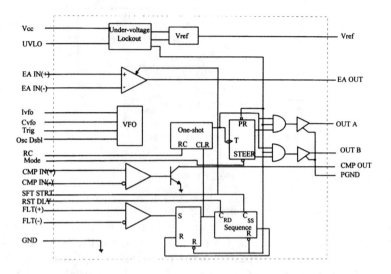

FIGURE 3–26 Functional block diagram of the UC1860 Resonant Mode Controller IC (Reproduced by permission of Unitrode Integrated Circuits)

3.6 Special DC to DC Converter Designs

3.6.1 Sub-5V Applications

As IC process lithography becomes finer and the need to reduce overall power dissipation becomes more important, ICs demand lower and better regulated power supply rails. Power supply rails for modern microprocessors and logic ICs have dropped below 5V to a 3.3V pseudo standard, and are fast heading towards sub-2V and sub-1V levels. These low-voltage supply requirements also differ in voltage level from system to system, and from version to version of the same IC.

Using Intel's Pentium as an example, some versions just require 3.3V while some require two supplies at 3.3 V and 3V. Yet other versions require supplies such as 3.383V and 3.525V, being the optimum voltages for maximum speed in the respective versions. Furthermore, the Pentium Pro comes with a 4-bit operating voltage code, which selects one of 16 discrete voltages between 2 to 3.5V (Travis 1996)–(Goodenough 1996).

Lower voltage levels however, do not translate to low power. These multi-million-transistor ICs use CMOS technology, which at today's high operating frequencies consume substantial amounts of current, though virtually no DC current. Power levels required are therefore rising.

Top-of-the line CPU power is approaching 100W, and desktop CPU power 30W. Systems such as high-performance multi-processor servers are approaching 400–600W. The combination of increasing power requirements and reducing voltage rails, requires the power source to provide and distribute very high currents within a system. For example, a 300W system running off a 3V supply must provide and distribute 100A of current.

Furthermore, clock rates heading towards 400–600 MHz imply high current transients as devices come out of the sleep mode within a clock cycle of a few nS.

In addition to microprocessors, there are high-speed data buses such as the 60MHz GTL (Gunning Transceiver Logic) bus, that are being recommended for interconnection of processors and peripherals on the mother board. It requires an active 1.2V or 1.5V terminator at each end, each potentially capable of handling up to 7A. Other data buses with similar requirements include Futurebus (2.1V) and Rambus (2.7V) (Goodenough 1995).

Therefore, the challenges facing the power supply designer in the sub 5-V range can be summarized as follows:

- Multiple voltage rails
- High efficiency while generating high currents (low power loss)
- Tight tolerances
- High accuracy
- High current transients

The basic design requirements in the face of these challenges are low power loss and higher gain and bandwidth in the control loop to handle tighter voltage regulation and transients.

3.6.2 Converters for Battery Operated Equipment and Battery Chargers

The proliferation of portable computers and hand-held communication devices has opened up another branch in DC/DC converter evolution. Users of these portable devices demand features such as compactness and lightness and low power consumption for extended periods of operation. Corresponding trends in DC/DC converters emphasize excellent conversion efficiency and compactness.

Additional constraints placed on the DC/DC converter include minimum noise intrusion into communications and audio circuits in the device. The increase in the integration of high-speed modems and CD ROM drives into portable computers has made power supply switching noise an important consideration.

A key requirement for designers of battery-powered products is that they minimize the number of cells used in the product. Products should ideally run off a few high-capacity cells to minimize size and cost. Tiny devices such as pagers run off one, or at most two, cells. So do telephones and PDAs. Converters operating off input sources as low as that of a single-cell alkaline battery (1V) are needed. Therefore, the voltage required is in most instances, greater than the voltage available from the cells. Furthermore, for extended battery life, the converter should be able to operate from waning batteries.

Some systems need to operate with input voltages that approach the output voltage. This low dropout condition requires the DC/DC converter duty cycle to approach 100 percent. Furthermore, a waning battery may swing the input voltage from a value above the output to one below.

High efficiency is also a prime consideration in battery operated equipment. It not only increases the operating time on a battery charge, but also reduces the heat which must be dissipated in or removed from the IC and the device. Low quiescent current is another requirement for extended battery charge.

Therefore, the challenges facing the power supply designer for battery powered applications can be summarized as follows:

- Low drop-out voltage
- Small size
- Extended use of battery charge
- Low EMI

3.6.3 Practical Design Approaches

This section describes some of the approaches taken in DC/DC converter and controller design to address the special requirements of sub-5V and battery powered applications. Illustrative examples of some successful solutions are presented.

3.6.3.1 Synchronous Rectification

In a conventional DC/DC converter such as the buck converter shown in Figure 3–5, typically an n-channel MOSFET with a low $R_{DS(on)}$ is selected for the switch. However, the diode's forward voltage becomes a limiting factor in improving the converter's efficiency as the output voltage drops.

This has lead to the design of synchronous converters, which replace the diode which is normally a Schottky, with another N-channel MOSFET. This is usually called the lower MOSFET, and the switch is called the upper MOSFET. The lower MOSFET conducts current during the off time of the upper MOSFET. Figure 3–27 shows a simplified diagram of a synchronous buck converter. Synchronous rectifiers are described further in Chapter 4 and (Sherman and Walters 1996).

In conventional synchronous converters, a single IC is used for PWM control and the synchronous drive of two external MOSFETs. Newer process technologies take another approach, where a *SynchroFET* integrates the two MOSFETS, their drive circuits and the synchronous control logic.

This IC can be used with a conventional PWM control IC to design a converter with superior features to those with discrete MOSFETs (Sherman and Walters 1996). Also, this approach allows the SynchroFET to be paired with many different PWM controllers to achieve various performance trade-offs.

HIP5015 and HIP5016 are a widely used pair of SynchroFETs from Harris Semiconductor. A PWM 5V to 3.3V DC/DC converter using a generic PWM controller and the HIP5015 SynchroFET is shown in Figure 3–28. This is typically used to derive 3.3 V from a 5V input. The implementation of a two-output converter using these SynchroFET ICs is described in (Goodenough 1996). The HIP5015/5016 is designed to run at over 1 MHz. The efficiency of a 5 to 3.3V converter running at

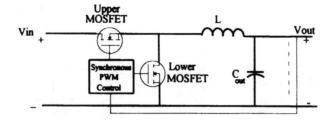

FIGURE 3-27 A synchronous buck converter

400kHz using this IC is reported to be 85 percent for load currents between 0.5 and 4.5A and over 90% for load currents between 0.8 and 2.8a (Goodenough 1996).

3.6.3.2 Increased Gain and Bandwidth Control Loop

In low voltage synchronous converters, the preferred method of control is the voltage mode. The power dissipated in the current sense resistor in current mode control cuts the efficiency by about 2 percent and is therefore unsuitable for low-voltage, high-efficiency applications. However, the low bandwidth of the voltage mode control loop reduces the converter's response to dynamic load conditions.

Maxim's three PWM controllers, the MAX 796/797/799, designed for portable computer and communication applications, are current mode controllers that drive synchronous rectifiers, primarily in the buck mode. Efficiencies as high as 97 percent are reported in.

Figure 3–29 shows the MAX 797 controller. The special features of this circuit are a proprietary PWM comparator for handling transients, a proprietary idle mode control scheme used at low-load conditions for extended battery life, and the reduction of PWM noise.

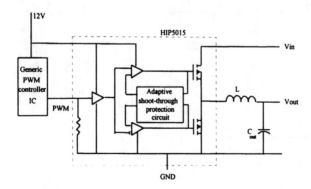

FIGURE 3-28 A PWM synchronous DC/DC converter using a SynchroFET (Copyright by Harris Corporation, reprinted by permission of Harris Semiconductor Sector)

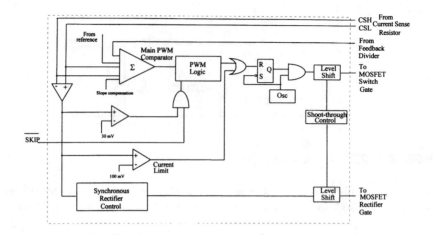

FIGURE 3-29 Block diagram of the MAX 797 PWM Controller (Reproduced with permission from Maxim Integrated Circuits Inc.)

This series is designed for output voltages as low as 2.5V, but can provide 1.5V with external circuitry. To provide the required DC accuracy at the low output voltage, while also handling the high-speed current transients, these processors require an external op amp in the error amplifier circuit, which operates as an integrator.

The heart of this PWM controller is a multi-input open-loop comparator that sums the output voltage error with respect to the reference, the current sense signal and a slope compensation ramp. This is of the direct summing type, and lacks the traditional error amplifier with the associated phase shift. This direct summing configuration approaches the ideal of direct cycle-by-cycle control of the output voltage. This PWM comparator is shown in Figure 3–30.

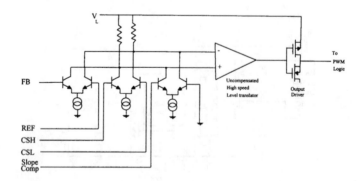

FIGURE 3-30 Block diagram of the main PWM comparator in the MAX 797 (Reproduced with permission from Maxim Integrated Circuits Inc.)

3.6.3.3 Idle Mode Control Scheme

The MAX797's approach for conservation of battery power makes use of the SKIP input shown in Figure 3–29. At light loads (SKIP = 0) the inductor current fails to exceed the 30 mV threshold set by the minimum current comparator. When this occurs, the minimum current comparator immediately resets the high-side latch at the beginning of the cycle unless the output voltage drops below the reference. This sends the controller into a variable frequency idle mode, skipping most of the oscillator pulses, in order to cut back gate-charge losses.

Operation at a fixed frequency regardless of load conditions is however advantageous in terms of lowering PWM noise interference. Steps can be taken to remove the known emissions at a fixed frequencies. Therefore, a low-noise mode is enabled in the MAX 797 by making SKIP high. This forces fixed frequency operation by disabling the minimum current comparator.

3.6.3.4 Updated Voltage Mode Control

A newer solution by Linear Technology employs an advance voltage mode control scheme in its LTC1430 controller. This controller, optimized for high power, 5V/3.xV applications operates at an efficiency greater than 90 percent in converter designs from 1A to greater than 50A output current. The LTC1430 uses a synchronous switching architecture with two N-channel output devices.

A block diagram of the LTC1430 is shown in Figure 3–31. The primary control loop is a conventional voltage mode feedback loop. The load voltage is sensed across the output capacitor by the SENSE– and SENSE+ inputs, divided and fed to the feedback amplifier, where it is compared to a 1.26V internal reference. The error signal is then compared to a saw tooth waveform from the oscillator to generate a pulse width modulated switching waveform.

Two other comparators in the feedback loop, MIN and MAX, provide high speed fault correction in situations where the feedback amplifier may not respond quickly enough. MIN compares the feedback signal to a voltage 40 mV (3 percent) below the reference, and MAX compares it to a voltage 3 percent above the reference.

If the output falls below the minimum level, the MIN comparator overrides the feedback comparator and forces the loop to full duty cycle (about 90 percent). If the output rises above the 3 percent level, the MAX comparator forces the duty cycle to zero. These two comparators prevent extreme output perturbations in the presence of fast output transients. Additionally, it senses output current across the drain source resistance of the upper N-channel FET, providing an adjustable current limit without an external sense resistor.

3.6.3.5 A Master/Slave Architecture

Power Trends' solution (Travis 1996) is a master/slave architecture for DC/DC converter modules which allows the increase of available current in 3A steps. The designs show a 5V/3.3V converter with a 3A or 8A master coupled with three 3A slaves. The slaves add 9A of current to the master's current output. This approach is an attractive solution which leaves room for future system upgrades.

90 POWER ELECTRONICS DESIGN HANDBOOK

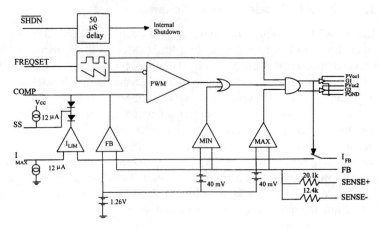

FIGURE 3-31 Block diagram of the LTC1430 Controller (Reproduced by permission of Linear Technology Corp)

3.6.3.6 Improved Process Technologies

Aiming at improved control loop bandwidth, Siliconix developed a high-speed CBiC/D (complementary bipolar/CMOS/DMOS) process which provides vertical pnp transistors with an f_t of 2 GHz. The Si9145 voltage mode controller built with this technology has an error amplifier with a functional closed-loop bandwidth of 100kHz, helping to handle transient loads provided by today's microprocessors (Goodenough 1995).

3.6.3.7 Switched Capacitor Converters

Switched capacitor (charge pump) converters use capacitors rather than inductors or transformers to store and transfer energy. The most compelling advantage of these converters is the absence of inductors. Compared with capacitors, inductors have greater component size, more EMI, greater layout sensitivity, and higher cost. Compared with other types of voltage converters, the switched capacitor converter can provide superior performance in applications that process low-level signals or require low noise operation.

Many capacitor-based voltage converters offer extremely low operating current—a useful feature in systems for which the load current is either uniformly low, or low most of the time. Thus, for small hand-held products, the light-load operating currents can be much more important than full-load efficiency in determining battery life. The basic operation of operation of switched capacitor voltage converters is shown in Figure 3-32 (Williams and Huffman 1988).

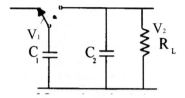

FIGURE 3-32   Principle of operation of switched capacitor converters

When the switch is in the left position, C_1 charges to V_1. The total charge on C_1 is given by, $q_1 = C_1 V_1$. When the switch moves to the right position, C_1 discharges to V_2. The total charge on C_1 now is given by, $q_2 = C_1 V_2$. the total charge transfer is given by,

$$q = q_1 - q_2 = C_1(V_1 - V_2).$$

If the switch is cycled at a frequency f, the charge transfer per second, or the current is given by,

$$I = fC_1(V_1 - V_2)$$ (3.6)
$$= \frac{(V_1 - V_2)}{R_{eq}}$$

where R_{eq} is given by $1/fC_1$.

The reservoir capacitor C_2 holds the output constant. The basic charge pump circuit can be modified for an inverted output by flipping C_1 in the internal switching arrangement. It can be used for doubling, tripling, or halving the input voltage with different external connections and by the use of external diodes. Figure 3–33 shows a simple switched capacitor IC implementation with on-chip switches.

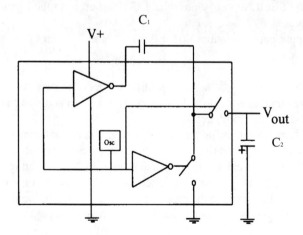

FIGURE 3-33   A simple IC switched capacitor implementation

The usage of switched capacitor converters in laptop computers and small hand-held communications devices is increasing as their output current capabilities are increasing and the supply current required by the portable devices is decreasing.

Simple switched capacitor converters such as the MAX660 can generate 100mA at 3.3V when powered from a 2-cell battery of alkaline, NiCd or NiMH cells or a single primary lithium cell. A disadvantage of a circuit such as this is the lack of regulation. This is overcome by adding a regulator externally.

Internal regulation in monolithic switched capacitor converter chips is achieved either as linear regulation or as charge pump modulation. Linear regulation offers low output noise.

Charge pump modulation controls the switch resistance and offers more output current for a given cost or size because of the absence of a series-pass resistor. Newer charge pump ICs employ an on-demand switching technique which enables low quiescent current and high output current capability at the same time. The simplicity of switched capacitor circuits, and hence its suitability for miniature equipment is amply illustrated in (Siliconix Inc. 1994), where the 8-pin Si7660 is used for voltage inversion, doubling and splitting.

3.6.3.8 SEPIC Converters

A relatively new converter architecture specially suited for battery operated equipment and battery charger applications is the SEPIC (Single Ended Primary Inductance Converter) topology. The SEPIC topology is shown in Figure 3–34.

The two inductors L_1 and L_2 are often two identical windings on the same core. This topology is essentially similar to a 1:1 transformer flyback converter, except for the addition of a capacitor C which forces identical AC voltages across the two inductors.

This is a step up/down topology with no inversion and no transformer. However, the SEPIC topology provides DC isolation from the input to the output due to the presence of the capacitor.

The input/output relationship for this converter is given by:

$$V_{out} = V_{in} D / (1 - D) \tag{3.7}$$

This topology has the advantage of being operable over wide input voltage range. Due to this, this topology finds applications commonly in battery powered equipment and in battery chargers.

A converter design using the LT1373 current mode switching regulator with a SEPIC converter for generating 3.3V from a single Li-Ion battery are described in (Essaff 1995). In a typical application such as this, the battery voltage at full charge is above the output voltage, and when discharged it is below the output voltage. This presents special difficulties in the converter design. The SEPIC architecture, with its wide input voltage range, is ideally suited for such applications.

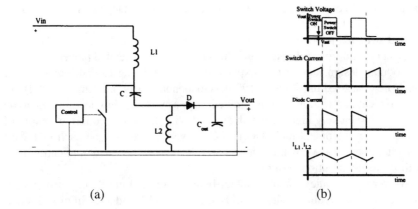

FIGURE 3-34 The SEPIC converter topology (a) The basic circuit (b) Associated waveforms

The use of the SEPIC topology in battery charging is illustrated in an application in. This topology allows charging even when the input voltage is lower than the battery voltage. Further, it allows the current sense circuit to be ground-referred and completely separated from the battery itself, simplifying battery switching and system grounding problems.

The LT1512 is a constant current controller for SEPIC converters, for charging NiCd and NiMH batteries. It can also provide a constant voltage source for charging lithium ion batteries [39]. A special feature of this controller is an internal low-dropout regulator which provides a 2.3V supply for all internal circuitry, allowing input voltages to vary from 2.7V to 25V. This enables charging of batteries from varied sources such as wall adapters, car batteries and solar cells.

3.7 DC to DC Converter Applications and ICs

The modularity of design, the wide-ranging power supply requirements, and the need for compactness, portability, and expandability of today's electronic equipment all contribute to the numerous applications that DC to DC converters have found in modern electronic equipment.

3.7.1 DC to DC Converter Applications

The area in which DC to DC converters have found the widest application is in distributed power systems. A distributed power system is one which uses many small regulated power supplies, each located as close as possible to the load (Goodenough 1995). The bulk supply develops and distributes via a bus, an arbitrary voltage level. At appropriate points DC to DC converters change this voltage to the levels needed for the local circuitry.

A distributed power system can save space and reduce the weight of the system. It can also improve reliability and the quality of the generated power, and facilitate modular design and system expansion. Issues relating to distributed power systems are discussed in (Ormond 1990), (Goodenough 1995), (Ormond 1992).

A typical application of a distributed power system is in telecommunication equipment. A relatively high voltage battery of 48 volts is located in the bottom of each electronic equipment rack. DC to DC converters are used on each rack card cage to step the voltage down to the required level. Another such application is in aircraft power distribution systems.

Modern electronic systems contain both analog and digital circuitry, where analog components require voltage levels such as 9, ±12, and ±15. In mixed mode logic circuits, the conventional 5V supply as well as a 3.3 V supply for the latest low-power ICs are required. In such complex cases, DC to DC converters perform the valuable function of generating all required voltages while saving cost and space. A common application of 5 to 12 V converters is in flash memory programmers which require a supply of 12 V.

Battery chargers that can be used off diverse power supplies ranging from the conventional mains supply to solar cells find the step-up/step-down ability of DC to DC converters a useful feature.

Another application area of DC to DC converters is in small, battery operated, portable equipment such as pagers, cameras, cellular phones, laptop computers, remote data acquisition and Instrumentation systems. The high efficiency and the small size of DC to DC converters become especially useful in such applications.

3.7.2 DC to DC Converter ICs

An increasing variety of chip-level DC to DC converters are appearing in the market. These devices are not only changing the way system designers structure their power supplies, but are also providing solutions to applications which previously required more costly, bulky and cumbersome approaches.

Integrated circuit DC to DC converters are available in a wide range of power and other capabilities. The lower power ICs can supply the exact voltage needed for a specific IC board, and higher power types can simplify the design by reducing the component count (Pryce 1988).

The key parameters and capabilities of modern low-power DC to DC converters are described in this section using a representative sample from principal manufacturers such as Maxim, Datel, Linear Technology Corp., etc. For example, output power capabilities range from less than 1 W to more than 300W. Single, double, or triple output configurations are available.

Common output voltages are 3.3V, 5, 12, and 15 volts, while input voltages are quite varied. Common devices perform 5 to 12V, 9 to 12V, and 5 to 3.3 V conversions. Table 3–3 shows key features of some representative DC to DC converters.

Linear Technology Corp. produces a wide range of DC to DC converters, ranging from 0.5W micropower devices to high efficiency 5A devices. Simple voltage doublers and dividers as well as circuits which can be used in many configurations are available. The LT1073, a gated oscillator mode IC, can operate from a supply of 1 V, and is typically used to generate 5 or 12 V from a single cell.

The XWR (Wide Input Range) Series by Datel Inc. are high efficiency current mode converters, which have typical efficiencies of about 85 percent. Typical applications of these are in telecommunications, automotive, avionic, and marine equipment, and in portable battery operated systems. In addition to the XWR series, Datel also has the LP series of converters having 3W, 4.5W, 5W, and 10W outputs.

Maxim Integrated Products have also developed a wide variety of DC to DC converters, operating in current mode and resonant mode. The MAX632 is a typical low power device, which can operate from an input voltage of 5V to produce a 12 V output. This type of converters is ideal for powering low power analog circuits from a 5V digital bus. They are relatively inexpensive and require very few external components.

The LM3578 from National Semiconductors is a low/medium power converter which can be used in buck, boost, and buck-boost configurations. This can supply output currents as high as 750 mA.

Siliconix's Si9100 and Si9102 can handle high input voltages. These chips are typically used in transformer coupled flyback and forward converter applications such as ISDN and PABX equipment and modems. Due to their high input voltage ratings, the Si9100 can operate directly from the -48V telephone line supply and the Si9102 from the -96 V double-battery telecom power supplies.

Further details of the above ICs as well as other similar products, application notes, etc., are found in (Linear Technology Corp. 1990)–(Essaff 1995).

3.8 State of the Art and Future Directions

As with all electronic disciplines, power supplies continue to get smaller, faster, better, and cheaper. The new techniques discussed in this chapter which extend the basic principles of DC/DC conversion give a taste of the future trends in power supply design and the driving forces behind them.

The adoption of progressively lower system voltages with progressively higher power in microprocessors drive DC/DC converter design in one direction, while battery operated systems provide another set of challenges. To face these challenges, and remain competitive in the market, key power supply designers continue to enhance their products using new material, new process technologies and new converter architectures.

TABLE 3-3 Some Representative DC to DC Converter ICs

Manufacturer	Model	Output Power (W)/Output Current (A)	Output Voltage (V)	Input Voltage (V)
Linear Technology Corp.	LT1070 LT1073 LT1074/76 LT 1173	5A 0.5 W 5/2 A 0.5W	Circuit Dependent 5, 12 5 5, 12	3 to 60 1 to 12 7 to 45/64 2 to 12 (step-down) 2 to 30 (step-up)
Datel Inc.	XWR Series	3, 10, 20W	3.3, ±5, ±12, ±15	4.6 to 13.2, 4.7 to 7, 9 to 18, 18 to 72
Maxim Integrated Products	MAX 632 MAX 742 MAX 756	0.025A 1A 0.3 A	12 5, 12, 15, 28, 3.3, 5	2 to 12.6 2 to 16.5 1.1 to 5.5
Siliconix Inc.	Si9100 Si9102 Si7660	0.350A 0.250A Up to 100 mA depending on V_{in}/V_{out}	$V_{out} < V_{in}$ $V_{out} < V_{in}$ $-V_{in}$, $2V_{in}$, $V_{in}/2$	10 to 70 10 to 120 1.5 to 10
National Semiconductors	LM3578	0.750A	$V_{out} < V_{in}$, $V_{out} < V_{in}$	2 to 40
Lambda	LSH6325P	2A	5 to 27	12 to 35
Vicor	Minimod	25 - 100W	5 to 48	12 to 300
Raytheon	RC4292	0.350A	-24 to 24	-20 to120

References

1. Maliniak, David. "Modern DC-DC Converter Sends Power Density Soaring." *Electronic Design*, August 21 1995, pp 59–63.
2. Travis, Bill. "Low-voltage power sources keep pace with plummeting logic and µP voltages." *EDN*, September 26, 1996.
3. Brown, Marty. *Practical Switching Power Supply Design*. Academic Press, Inc., 1990.
4. Brown, Marty. *Power Supply Cookbook*. Butterworth-Heinemann, 1994.
5. Bose, B. K. (Editor). *Modern Power Electronics: Evolution, Technology and Applications*. IEEE Press, 1992.

DC to DC Converters 97

6. Chryssis, George C. *High Frequency Switching Power Supplies (Second Edition).* McGraw-Hill, 1989.
7. Cuk, S. and R. D. Middlebrook. "A New Optimum Technology Switching DC-to-DC Converter." IEEE Power Electronics Specialist Conference Record, June 14–16, 1994, pp 160–179.
8. Moore, Bruce D. "Step-up / Step-down Converters Power Small Portable Systems." *EDN*, February 3 1994, pp 79–84.
9. Linear Technology Corporation. "1992 Databook Supplement." 1992.
10. Linear Technology Corporation. *1990 Linear Databook.* 1990.
11. Datel Inc. "Datel Databook Volume 4: Power." 1991.
12. Siliconix Inc. "Power Products Data Book." 1994.
13. Maxim Integrated Products. "Maxim 1994 New Releases Data Book." Vol. 3, 1994.
14. Motorola Inc. "Linear/Switchmode Voltage Regulator Handbook." 1987.
15. Unitrode Integrated Circuits. "Product & Application Handbook." 1995–96.
16. Steigerwald, Robert L. "High Frequency Resonant Transistor DC-DC Convertors." IEEE Transactions on Industrial Electronics, May 1984, pp 181–191.
17. Cuk, S. and Dragan Maksimovic. "Quasi-Resonant Cuk DC-to-DC Converter Employs Integrated Magnetics - Part I." *PCIM*, December 1988, pp 16–24.
18. Wofford, Larry. "A New Family of Integrated Circuits to Control Resonant Mode Power Converters." *PCIM,* April 1990, pp 26–35.
19. Goodenough, Frank. "Power supply rails plummet and proliferate." *Electronic Design,* July 24, 1995, pp 51–55.
20. Goodenough, Frank. "Fast LDOs and switchers provide sub–5V power." *Electronic Design,* September 5, 1995, pp 65–74.
21. Goodenough, Frank. "Driver ICs create 1 or 2 processor power rails." *Electronic Design,* September 16, 1996, pp 65–74.
22. Sherman, Jeffrey D. and Michael M. Walters. "Synchronous rectification: improving the efficiency of buck converters." *EDN,* March 14, 1996.
23. "Step Down Controllers with Synchronous Rectifier for CPU Power." http://www.maximic.com/PDF/1190.pdf.
24. "LTC1430 High Power Step-Down Switching Regulator Controller." http://www.linear.com/cgibin/database?function=elementinhtml&filename =DataSheet.html&name=DataSheet&num=22
25. Williams, Jim and Brian Huffman. "Switched Capacitor Networks Simplify DC/DC Converter Design." *EDN,* November 24, 1988, pp 171–175.
26. Siliconix Inc. "Analog Integrated Circuits Data Book." 1994.
27. Essaff, Bob. "250kHz, 1mAIQ Constant requency Switcher Tames Portable Systems Power." *Linear Technology Design Note 108,* July 1995.
28. LT1512 SEPIC Constant Current/Constant Voltage Battery Charger. http://www.lineartech.com/pdf/1512f.pdf.
29. Ormond, Tom. "Distributed Power Schemes Simplify Design Tasks." *EDN,* December 6, 1990, pp 132–136.
30. Goodenough, Frank. "System Designers vs. Power Supplies." *Electronic Design,* May 15, 1995, pp 22.
31. Ormond, Tom. "Distributed Power Schemes Put Power Where You Need It." *EDN,* July 6, 1992, pp 158–164.
32. Pryce, Dave. "Growing Array of 1-Chip DC/DC Converters Provides Power For Diverse Applications." *EDN,* February 18, 1988, pp 73–80.
33. Sherman, Leonard H. "DC/DC Converters adapt to the needs of low-power circuits." *EDN,* January 7, 1988.

34. Pflasterer, Jim. "DC-DC Converter ICs Address the Needs of Battery Powered Systems." *PCIM*, April 1997.
35. "Trends in Battery Power Recharge DC/DC Converter Advances." *AEI*, February 1997.

Bibliography

1. Mitchell, Daniel M. *DC-DC Switching Regulator Analysis*. McGraw-Hill Inc., 1988.
2. Cuk, S. "Basics of Switched-Mode Power Conversion: Topologies, Magnetics and Control." *Powerconversion International,* August, September, October 1981.
3. Lee, Fred C., Wojciech A. Tabisz, and Milan M. Jovanovic. "Recent Developments in High Frequency Quasi-Resonant and Multi-Resonant Converter Technologies." Proceedings of the 1988 European Power Electronic Conference, 1989, pp 401–410.
4. Williams, Jim and Brian Huffman. "Precise Converter Designs Enhance System Performance - Part I." *EDN*, October 13 1988, pp 175–185.
5. Strassberg, Dan. "A Surfeit of Power Supply Voltages Plagues Designs of Compact Products." *EDN*, May 12, 1994, pp 55–62.

CHAPTER **4**

Off-the-Line Switchmode Power Supplies

4.1 Introduction

The basic theory of the switching power supply has been known since the 1930s. Practical switchmode power supplies (SMPS) have existed since the 60s, starting with components designed for other uses and thus poorly characterized for SMPS use. Early SMPS systems were running at very low frequencies, usually between a few KHz to about 20 KHz. Early SMPS designs justified the SMPS techniques for higher output power ratings, usually above 500 watts.

Towards the mid 80s most SMPS units were available for much lower power ratings with multiple output rails, while oscillator frequencies were reaching the upper limit of 100 KHz. Today most switchers operate well above 500 KHz and newer techniques allow the systems to run at several MHz, making use of new magnetics, surface mount components, and the resonant conversion techniques.

The rapid advancement of microelectronics in the last two decades (1975 to 1995) has created a necessity for the development of sophisticated, efficient, light weight power supplies which have a high power-to-volume (W/in^3) ratio with no compromise in performance. The high-frequency switching power supply meets these demands.

Recently it has become the prime powering source in the majority of modern electronic systems. The trends associated with the switchmode power supplies for the electronic products and systems are (a) to reach direct off-the-line design approach, (b) higher frequencies be utilized, (c) output rating/volume being increased and minimizing the components and increasing reliability.

This chapter discusses the basic blocks of direct off line SMPS units and the newer trends in technology such as distributed DC power, etc. Configurations used for switching blocks are not discussed here as they are covered in the previous chapter. Introduction to magnetic components, use of magnetic amplifiers, and the modern trends in high frequency capacitors, etc. are also introduced.

4.2 Building Blocks of a Typical High Frequency Off-the-Line Switching Power Supply

The building blocks of a typical high-frequency off-the-line switching power supply are depicted in Figure 4–1(a). The basic operation of the switching power supply was described in the previous chapter. This block diagram shows other important sections such as the AC line RFI filter block, the ancillary and supervisory blocks, and the I/O isolation block. The EMI/RFI filter could be either part of the power supply or external to it, and is generally designed to comply to national or international specifications, such as the FCC class A or class B and VDE-0871.

Within the past few years, due to emphasis placed on the power quality issues, power factor correction as applicable to the nonlinear behavior of the input current waveform of a switcher has become an important issue. For power supplies with output capacity over 700VA power factor correction is becoming compulsory and Chapter 9 provides details related to power factor correction techniques. Figure 4–1(b) shows a block diagram of a SMPS with power factor correction.

The ancillary and supervisory circuits are used to protect the power supply as well as the load from fault conditions. Generally, each power supply has current-limit protection to prevent its destruction during overload conditions. Overvoltage protection is part of the supervisory circuits used to protect the load from power supply failures. It is important to note that although in a linear power supply, overvoltage conditions were common during the short circuits of the series-element, in a switching power supply, failure of the switching element normally results in a non-output condition. However, the output of the switcher will go high if the feedback loop is opened.

Input/output isolation is essential to an off-the-line switcher. The isolation may be optical or magnetic, and it should be designed to comply to Underwriters Laboratories (UL), Canadian Standards Association (CSA), Verband Deutscher Electronotechniker (VDE), or International Electrotechnical Commission (IEC) safety standards. Thus the UL and CSA require 1000 V AC isolation voltage withstand while VDE and IEC require 3750V AC. Consequently, the power transformer has to be designed to the same safety isolation requirements also.

4.2.1 The Input Section

As mentioned previously, an off-the-line switching power supply rectifies the AC line directly without requiring a low-frequency line isolation transformer between the AC mains and the rectifiers. Since in most of today's electronic equipment the manufacturers are generally addressing an international market, the power

Off-the-Line Switchmode Power Supplies **101**

(a)
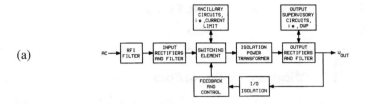

(b)
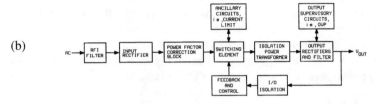

FIGURE 4-1 The building blocks of a typical off-the-line high-frequency switching power supply (a) Without power factor correction (b) With power factor correction

supply designer must use an input circuit capable of accepting many different line voltages, normally 90 to 130V AC or 180 to 260V AC. Figure 4–2 shows a realization of such a circuit by using the voltage doubler technique. When the switch is closed, the circuit may be operated at a nominal line of 110V AC.

During the positive half cycle of the AC, capacitor C_1 is charged to the corresponding peak voltage, approximately 155V DC, through diode D_1. During the negative half cycle, capacitor C_2 is charged to 155V DC through diode D_4. Thus, the resulting DC output will be the sum of the voltages across $C_1 + C_2$, or 310 V DC. When the switch is open, D_1 to D_4 form a full-bridge rectifier capable of rectifying a nominal 220-V AC line and producing the same 310V DC output voltage.

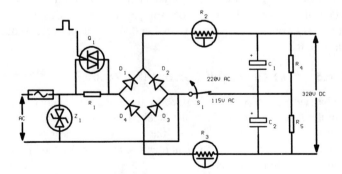

FIGURE 4-2 Input section of an SMPS showing basic components.

4.2.1.1 Selection of Basic Components

In selecting the primary components such as the bridge rectifier (or the diodes) and the filter capacitors, the designer must look for several important specifications.

4.2.1.2 Diodes and Capacitors

For diodes, maximum forward rectification current capability, Peak Inverse Voltage (PIV) capability, and the surge current capability (to withstand the peak current associated with turn-on) are the most important specifications.

Proper calculation and selection of the input rectifier filter capacitors is very important, since this will influence the performance parameters such as low-frequency AC ripple at the output power supply and the holdover time. Normally high-grade electrolytic capacitors with high ripple current capacity and low Equivalent Series Resistance (ESR) are used with a working voltage of 200 V DC minimum. Resistors R_4 and R_5, shown in Figure 4–2 shunting the capacitors, provide a discharge path when the supply is switched off.

4.2.1.3 Fuses

Modern fuse technology is an advanced science. New and better fuses are continually being developed to meet the more demanding requirements for protection of semiconductor circuitry. To obtain the most reliable long-term performance and best protection, a fuse must be knowledgeably chosen to suit the application. Fuses are categorized by three major parameters; current rating, voltage rating, and, most important, "let-through" current, or I^2t rating.

4.2.1.4 Current and Voltage Rating

Current rating of a fuse is the rms value or the maximum DC value that must exceed before blowing the current demanded by the protected circuit. The voltage rating of a fuse is not necessarily linked to the supply voltage. Rather, the fuse voltage rating is an indication of the fuse's ability to extinguish the arc that is generated as the fuse element melts under fault conditions.

The voltage across the fuse element under these conditions depends on the supply voltage and the type of circuit. For example, a fuse in series with an inductive circuit may see voltages several times greater than the supply voltage during the clearance transient.

4.2.1.5 "Let-Through" Current (I^2t Rating)

This characteristic of a fuse is defined by the amount of energy that must be dissipated in the fuse element to cause it to melt. This is sometimes referred to as the pre-arcing let-through current. To melt the fuse element, heat energy must be dissipated in the element more rapidly than it can be conducted away. This requires a defined current and time product.

The heat energy dissipated in the fuse element is in the form of watt-seconds (joules), or $I^2.Rt$ for a particular fuse. Since the fuse resistance is a constant, this is proportional to I^2t, normally referred to as the I^2t rating for a particular fuse or the pre-arcing energy. The I^2t rating categorizes fuses into the more familiar "slow-blow," "normal," and "fast-blow" types. It should be noted that the I^2t energy can be as much as two decades greater in a slow-blow fuse of the same DC current rating. For example, a 10A fuse can have an I^2t rating ranging from 5 $A^2.s$ for a fast fuse to 3000 $A^2.s$ for a slow fuse. Selection of fuse ratings for off line SMPS are discussed in Billings (1989).

For high power semiconductors manufacturers indicate a value of I^2t, for 10 ms (for 50Hz) or 8.3 ms (for 60 Hz), that should not be exceeded. Comparing this value with the fuse I^2t permits us to verify the protection. De Palama and Deshayes (1996) provide details of protecting power semiconductors using the appropriate fuses.

4.2.2 Input Protective Devices

4.2.2.1 Inrush Current

An off-the-line switching power supply may develop extremely high peak inrush currents during turn-on, unless the designer incorporates some form of current limiting in the input section. These currents are caused by the charging of the filter capacitors, which at turn-on present a very low impedance to the AC lines, generally only their ESR. If no protection is employed these surge currents may approach hundreds of amperes.

Two methods are widely employed in introducing an impedance to the AC line at turn-on and in limiting the inrush current to a safe value. One is using a resistor-triac arrangement, and the other using negative temperature coefficient (NTC) thermistors. Figure 4–2 shows how these elements R_2 and R_3 may be employed in a power supply.

4.2.2.2 The Resistor-Triac Technique

Using this inrush current limiting technique, a resistor is placed in series with the AC line. The resistor is shunted by a triac which shorts the resistor out when the input filter capacitors have been fully charged. This arrangement requires a trigger circuit which will fire the triac on when some predetermined conditions have been met. Care must be taken in choosing and heat-sinking the triac so that it can handle the full input current when it is turned on.

4.2.2.3 The Thermistor Technique

This method uses NTC thermistors placed on either the AC lines or the DC buses after the bridge rectifiers, as shown in Figure 4–2. When the power supply is switched on the resistance of the thermistor(s) is essentially the only impedance across the AC line, thus limiting the inrush current.

As the capacitors begin to charge, current starts to flow through the thermistors heating them up. Because of their negative temperature coefficient, as the thermistors heat up their resistance drops. When the thermistors are properly chosen, their resistance at steady-state load current will be a minimum, thus not affecting the overall efficiency of the power supply.

4.2.2.4 Input Transient Voltage Protection

Although the AC mains are nominally between 110V to 240V AC, it is common for high-voltage spikes to be caused by nearby inductive switching or natural causes such as electrical storms or lightning. On the other hand, inductive switching voltage spikes may have an energy content:

$$W = \frac{1}{2} \cdot LI^2 \tag{4.1}$$

Where L is the leakage inductance of the inductor, and I is the current flowing through the winding. Therefore, although these voltage spikes may be short in duration, they may carry enough energy to prove fatal for the input rectifiers and the switching transistors, unless they are successfully suppressed.

The most common suppression device used in this situation is the metal oxide varistor (MOV) type transient voltage suppresser, and it may be used as shown in Figure 4–2 across the AC line input. This device acts as a variable impedance; that is, when a voltage transient appears across the varistor, its impedance sharply decreases to a low value, clamping the input voltage to a safe level. The energy in the transient is dissipated in the varistor. In more modern designs solid state devices and MOVS are mixed to form more guaranteed protection. Chapter 6 provides details.

4.2.3 RFI/EMI Filter

Every switching power supply is a source of Radio Frequency Interference (RFI) generation because of the very fast rise and fall times of the current and voltage waveforms inherent in the DC-DC converter operation. The main sources of switching noise are the switching transistor(s), the mains rectifier, the output diodes, the protective diode for the transistor, and of course the control unit itself. Depending upon the topology of the converter used, the RFI noise level at the mains input may vary.

Flyback converters, which by design have a triangular input current waveform, generate less conducted RFI noise than converters with rectangular input current waveforms, such as forward or bridge converters. Fourrier analysis shows that the amplitudes of the high-frequency harmonics of a triangular current waveform drop at a rate of 40 decibels per decade, compared to a 20 decibel per decade drop for a comparable rectangular current waveform.

United States and international standards for EMI-RFI have been established which require the manufacturers of electronic equipment to minimize the radiated and conducted interference of their equipment to acceptable levels. In the United

States the guiding document is the Federal Communications Commission (FCC) Docket 20780, while internationally the West German Verband Deutscher Elektronotechniker (VDE) safety standards have been widely accepted. It is very important to understand that both the FCC and VDE standards exclude subassemblies from compliance to the rule; rather, the final equipment, where the switching power supply is used, must comply with the EMI-RFI specifications.

Both the FCC and VDE are concerned with the suppression of RFI generated by equipment connected to the AC mains employing high-frequency digital circuitry. The VDE has subdivided its RFI regulations into two categories, the first being unintentional high-frequency generation by equipment with rated frequencies from 0–10 kHz, i.e., VDE-0875 and VDE-0879, and the second dealing with intentional high-frequency generation by equipment using frequencies above 10 kHz, i.e., VDE-0871 and VDE-0872. The FCC on the other hand includes in its RFI regulations all electronic devices and systems which generate and use timing signals or pulses at a rate greater than 10 kHz. Chryssis (1989) summarizes the FCC and VDE RFI requirements.

Figure 4–3 shows a differential noise filter and a common-mode noise filter in series. In many cases the common-mode filter is used alone, as it can often eliminate as much as 90 percent of the unwanted noise.

In EMI/RFI filters, the cores used for the common mode filter are mainly high permeability toroids. Here, the cores are wound either bifilar, using both of the line current wires, or two identical windings on opposing sides of a toroid. In this way the flux created by each line current cancels the flux from the other line so the high line current does not saturate the core.

Because of this arrangement, any unwanted noise that attempts to flow out of the power supply through both input leads in the same phase and same direction will be attenuated by the high impedance of the core. Ferrite toroids with permeabilities of 5,000 (J Material), 10,000 (W material), and 15,000 (H material) are popular because of their high permeability or impedance characteristics.

In-line filters are necessary to eliminate noises that are in phase with the line currents. Heavy line currents flowing through the windings of these filters tend to saturate the core. Therefore, the cores must be gapped; preferred types are ferrites and powder cores.

In the RFI/EMI filter blocks the capacitors bypass the high frequency components generated from the switching block due to low effective impedance at the high frequencies. Magnetics Inc. (1995) provides design details of RFI/EMI filters.

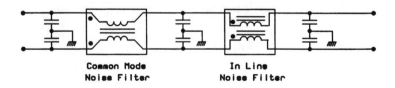

FIGURE 4–3 EMI/RFI filter

4.3 Magnetic Components

4.3.1 Transformers and Inductors

Magnetic elements are the cornerstone of all switching power supply designs but are also the least understood. There are three types of magnetic components inside switching power supplies: a forward-mode transformer or a flyback-mode transformer, an AC filter inductor, and a DC filter inductor or a magnetic amplifier, etc.

Each has its own design approach. Although the design of each of these magnetic components can be approached in an organized step-by-step fashion, it is beyond the intent of this chapter. For further information Brown (1990) is suggested.

4.3.1.1 Magnetic Materials

Magnetic materials can be divided into three broad categories, each with its own distinct range of characteristics and capabilities:

i. Magnetic metals and alloys, in tape or laminated form (e.g., silicon steel, Mumetal, Permalloy, Amorphous metals);
ii. Powdered magnetic metals (e.g., "Powdered Iron," Moly Permalloy Power); and
iii. Magnetic ceramics, or ferrites.

4.3.1.1.1 Magnetic Metals

Typically these have a high saturation flux density (7 to 23KG) and a high permeability (1K to 200K). Electrical conductivity is also high, requiring their use in laminated or tape form to reduce eddy current loss. A large range of shapes is available, limited primarily to "two dimensional" shapes by the wound tape or laminated construction.

The high flux capability makes them most suitable as line frequency (50 to 400 Hz) transformers and inductors. Eddy current losses limit use at high frequency, although some find application up to several hundred KHz.

4.3.1.1.2 Powdered Metals

These are magnetic metals which are powdered, coated with an insulator, and pressed and sintered to shape. Eddy currents are reduced by the small particle size, extending the useful frequency range to a few megahertz in some applications.

The insulated particles create a distributed air gap, giving these materials a low permeability, typically in the range of 8 to 80, with values to 550 obtainable in MPP. The low permeability and a tolerance for DC current makes these materials suitable for filter inductors at medium frequencies (typically 1 to 100 KHz). A relatively stable and well defined permeability allows them to be used for nonadjustable tuning inductors, with a typical tolerance of a few percent. Saturation flux densities are typically in the range of 5 to 10 KG, but a very gradual saturation characteristic limits the maximum flux when inductance variations must be minimized.

Available core shapes tend to be limited; toroids are the most common, with "E" cores, pot and cup cores, and a few similar shapes also available in some materials.

4.3.1.1.3 Ferrites

Ferrites are dark grey or black ceramic materials. They are very hard, brittle, and chemically inert. Most common ferrite materials such as MnZn and NiZn exhibit good magnetic properties below Curie Temperature (T_c). Ferrites are characterized by very high resistivities, making them the most suitable material for transformers and inductors operating at high frequencies (up to tens of MHZ) when low loss or high Q is of major concern. Saturation flux densities are low, in the 2 to 5 KG range, with permeabilities in the middle range of a few hundred to 15,000.

Ferrites have the widest range of available core shapes, including any of the shapes for the other materials. Maximum linear dimensions are limited to a few inches by standard fabrication processes, although large cores can be assembled in sections.

The flexibility in shape often makes ferrites the material of choice, even when their other desirable properties are not required. They can easily be magnetized and have a rather high intrinsic resistivity. These materials can be used up to very high frequencies without laminating as is the normal requirement for magnetic metals.

NiZn ferrites have a very high resistivity and are therefore most suitable for frequencies over 1 MHz but MnZn ferrites exhibit higher permeabilities (μ_i) and saturation induction levels (B_s). Different kinds of power ferrite material are available from many vendors. For example available power ferrites from Phillips are available as per data in Table 4–4.

Figure 4–4 indicates the typical block diagram of a SMPS with magnetic components. Table 4–1 indicates the core types and magnetic materials usable in each block.

A comparison of the characteristics of core material, advantages, and disadvantages are described in Magnetics Inc. (1995).

TABLE 4–1 Usable Magnetic Materials in SMPS Blocks (Courtesy of Magnetics Inc., USA)

Core Material and Type	Common Mode Filter	In-line Filter	PFC Inductor	Output Transformer	Magnetic Amplifier	Filter Inductor
Ferrite toroids	yes			yes		
Ferrite shapes (ungapped)	yes			yes		
Ferrite pot cores				yes		
Ferrite shapes (gapped)				yes		
MPP		yes	yes			yes
50Ni-50Fe powder cores		yes	yes			yes
Gapped ferrites		yes	yes			yes
Powdered iron		yes	yes			yes
SiFe laminations		yes				yes
Kool Mµ Powder cores		yes	yes			yes
Ultra Mµ Amorphous tape cores	yes					
Ni-Fe tape wound cores					yes	
Amorphous tape wound cores				yes		
Cut core Ni-Fe/Amorphous				yes		yes
Ni-Fe laminations				yes		
Cobalt-base amorphous tape wound cores					yes	
Square loop ferrite toroids					yes	

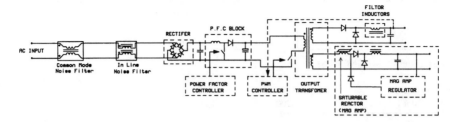

FIGURE 4-4 Typical SMPS blocks indicating the use of magnetic components

Table 4-2 compares the core materials.

4.3.1.2 Core Geometry and Common Core Materials

Cores come in many shapes and sizes. The most common core types are shown in Figure 4-5. There are many more types, but they are all based upon these basic styles. Some of the important considerations when selecting a core type are core material, cost, the output power of the power supply, the physical volume the transformer or inductor must fit within, and the amount of RFI shielding the core must provide. Table 4-3 provides ferrite core geometry considerations.

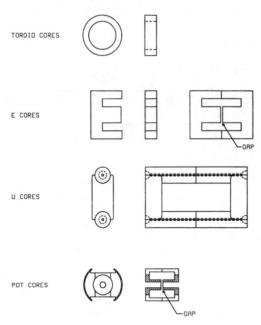

FIGURE 4-5 Common core types

TABLE 4-2 Core Material Considerations (Courtesy of Magnetics, Inc., USA)

	Flux Density	Initial Perm.	Frequency* Range
Ferrite Toroids MAGNETICS™			to >MHz
J Mat'l	4300	5000	
W Mat'l	4300	10,000	
H Mat'l	4200	15,000	
Ferrite Shapes			
K Mat'l	4600	1500	to 2MHz
R Mat'l	5000	2300	to 200KHz
P Mat'l	5100	2700	to 100KHz
F Mat'l	4700	3000	to 100KHz
MPP Cores	7000	14-550	<1MHz
50 Ni-50 Fe Powder Cores	15,000	60-200	<1MHz
KOOL Mµ® Powder Cores	11,000	60-125	<1MHz
Powdered Iron	9000	22-90	<1MHz
Silicon-Fe Laminations	16,000	4000	<1000Hz
Ni/Fe Tape Cores, Ni/Fe Bobbin Cores	7,000 to 15,000	to 100,000	to 100KHZ
Amorphous Tape Cores (cobalt-base)	5,000	to 100,000	to 500KHz
Amorphous Tape Cores (iron-base)	16,000	10,000	to 500KHz
Si-Fe Tape Cores	16,000	4000	<1000Hz
Ni-Fe Cut Cores	15,000	15,000	to 100KHz

TABLE 4-3 Ferrite Core Geometry Considerations

	Core Cost	Bobbin Cost	Winding Cost
Pot Core	high	low	low
Slab-sided Core	high	low	low
E Core	low	low	low
Ec Core	medium	medium	low
Toroid	very low	none	high
PQ Core	high	high	low

TABLE 4–2 Continued

Max. op. Temp.	Core Losses	Core Cost	Winding Cost	Temp. Stability	Mounting Flexibility
100°C	lowest	low	high	fair	fair
125°C		(see Table 4–3 below for geometry considerations)			
125°C	(1)				
125°C	(2)				
125°C	(3)				
200°C	low	high	high	good	fair
200°C	low	high	high	good	fair
200°C	low	low	high	good	fair
200°C	high	lowest	high	fair	fair
300°C	highest	low	low	fair	good
200°C	low to medium	high	high	good	fair
150°C	low	high	high	good	fair
150°C	low	high	high	good	fair
300°C	highest	medium	high	good	fair
150°C	medium	high	low	good	fair

TABLE 4–3 Continued

Winding Flexibility	Assembly	Mounting Flexibility	Heat Dissipation	Shielding
good	simple	good	poor	excellent
good	simple	good	good	good
excellent	simple	good	excellent	poor
excellent	medium	fair	good	poor
fair	none	poor	good	good
good	simple	fair	good	fair

For modern switching power supplies some of the commonly used core materials are F, K, N, R, and P materials from Magnetics Inc.; 3C8, 3C85 from FerroxCube Inc; or H7C4 and H7C40 materials from TDK. These ferrite materials offer the low core losses at the operating frequencies between 80KHz to 2.0 MHz. Table 4–4 indicates the available power ferrite materials from Phillips, their usage, and characteristics.

Resonant power supplies make it possible to exploit the area beyond 1 MHz, which requires power ferrite transformers with useful properties in the 1–3 MHz frequency range. Meeting this need is a new power transformer core material, 3F4, MnZn ferrite. Precise control of the fine-grained microstructure, which requires both a high purity powder processing and an adapted sintering technology, enables the 3F4 material to obtain these properties. The chemical composition includes dopants to reduce the hysterisis loss and high frequency wall damping loss. The material also suppresses electrical conductivity and thus the eddy current losses. For further details Visser and Shpilman (1991) and Bates (1992) are suggested.

TABLE 4–4 Power Ferrite Materials from Phillips, Their Usage and Frequency Characteristics

Material Grade	Characteristics	Usage	Available core types
3B8	Medium Frequency (<200 kHz)	small power transformers and general purpose transformers	RM, P
3C80	Low Frequency (<100 kHz)	Power transformers in TV applications	E, EF, ETD, EC, U
3C10	Low Frequency with improved saturation level	Flyback transformers	U
3C85	Medium Frequency (<200kHz)	Industrial use	E, EF, ETD, EC, RMP, EP, Ring Core
3F3	High Frequency (up to 1MHz)	Resonant Power supplies	E, EF, ETD, RM, P, EP, Ring Core
3F4	High Frequency (1-3 MHz)	Resonant power supplies	

4.3.1.3 Winding Techniques and Practical Considerations

The design and the winding technique used in the magnetic component's design has a great bearing on the reliability of the overall power supply. Two situations arise from a poor transformer design; high voltage spikes are generated by the rate of transitions in current within the switching supply, and the possibility of core saturation can arise during an abnormal operational mode.

Voltage spikes are caused by a physically "loose" winding construction of a transformer. The power supply depends upon the quick transmission of transitions in current and voltage between the transformer windings. When the windings are physically wound distant from one another, the leakage inductances store and release a portion of the energy supplied to a winding in the form of voltage spikes. It also delays the other windings from seeing the transition in the drive winding. Spikes can cause the semiconductors to enter avalanche breakdown and the part can instantly fail if enough energy is applied. It can also cause significant RFI problems.

A snubber is usually the solution, but this lowers the efficiency of the power supply. Core saturation occurs when there are too few turns on a transformer or inductor. This causes the flux density to be too high and at high input voltages or long pulse widths, the core can enter saturation. Saturation is when the core's cross sectional area can no longer support additional lines of flux. This causes the permeability of the core to drop, and the inductance value to drop drastically.

4.3.1.4 Transformers

For forward mode operation transformers may use either tape wound cores or ferrite cores. Tape wound cores are used at the lower frequencies because their high flux density minimizes the transformer size. As the frequency increases, high frequency materials are required and the tape thickness must be reduced, both of which increase the core cost. This makes the tape wound core more costly than ferrites at frequencies over 10 KHz.

Tape wound cores often have to be gapped with a minimal type air gap. Figure 4–6 shows the hysterisis loop of a typical tape wound core both with and without an air gap. Because the loop is extremely square, any unbalance in the switching process could cause saturation for some part of the cycle, thus generating extra losses in the circuit and distorting the output. As indicated in Figure 4–6(a), the gapped core will require a large difference in transistor currents before it will saturate the core.

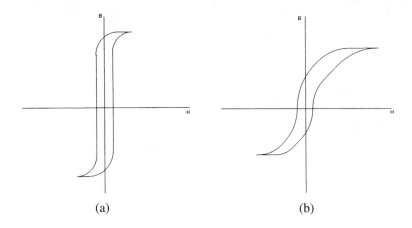

FIGURE 4–6 B-H loop for a tape wound core (a) Ungapped (b) Gapped

One other method to eliminate this problem of core saturation is to use a composite core. These cores are made with an ungapped tape wound core surrounded by another tape wound core containing a small air gap. Composite cores avoid the problems of unbalanced transistors and also reduce or eliminate spikes in the output. The state of the art power supplies running at frequencies over 200 KHz, where the losses could be high, ferrites are the dominant core material because of their lower cost and availability of variety of shapes and sizes.

4.3.1.5 Flyback Transformers

The transformer cores for flyback circuits require an air gap in the core so that they do not saturate from the DC current flowing in their windings. As mentioned above, the tape wound core and ferrite core can both be provided with air gaps. The toroidal Permalloy powder core is also used because it offers a distributed-type air gap. The frequency of the unit will dictate the selection of either tape wound cores or ferrites as described previously, while the powder cores will operate at frequencies up to 50 KHz. Operation of the flyback converter is based on the storage of energy in the core and its air gap during the on time of the switch and discharging the energy into the load during the switch's off time. The magnetic core operates in one quadrant of its B-H curve. For high energy transfer in a small volume, the core should have the B-H curve shown in Figure 4–7. An ideal core has a large available flux swing (ΔB), low core losses, relatively low effective permeability, and low cost. Moly Permalloy Powder (MPP) cores, gapped ferrites, Kool Mμ, and powdered iron cores are used in this application.

In the case of ferrites, to increase their energy transfer ability ferrites must have an air gap added into the core's structure (Figure 4–8). Although the air gap increases the energy transfer of the ferrite core, it can create EMI problems. Better shielded core shapes, such as pot cores, PQ cores, RM cores, and EP cores, are used to combat stray magnetic flux emitted from the air gap. The disadvantage of these shapes is higher cost.

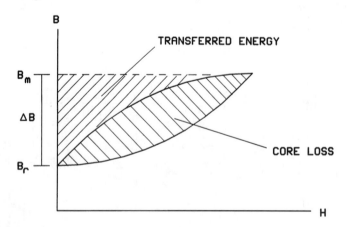

FIGURE 4–7   B-H curve for flyback power supply

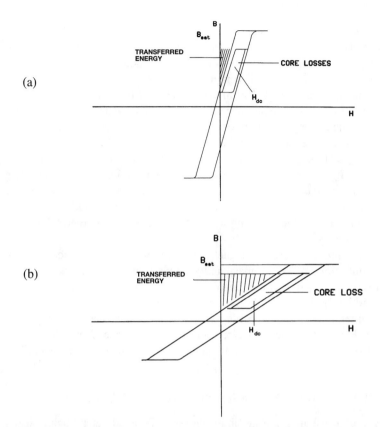

FIGURE 4-8 Effect of gap in ferrite core (a) Ungapped ferrite (b) Gapped ferrite

4.3.1.6 Inductors

The switching regulator usually includes a power inductor. Because of the large DC current through its windings, the core must have a large air gap to keep it from saturating. Cores used here are gapped ferrite cores, Permalloy powder cores, and high flux powder cores.

In the current wave form, because there is usually only a small ripple, the AC flux swing in the core is small. Therefore, powder iron cores and silicon laminations may also be used. They are less expensive than the above mentioned cores but have much higher losses, and, therefore, care must be taken in the design of such units to avoid overheated inductors and possible damage that may be caused to other components.

Ferrite E cores and pot cores offer the advantages of decreased cost and low core losses at high frequencies. For switching regulators, F or P materials are recommended because of their temperature and DC bias characteristics. By adding air gaps to these ferrite shapes, the cores can be used efficiently while avoiding saturation.

4.3.2 Planar Magnetics

The concept of planar magnetics has existed for many years, however, the idea has recently gained popularity due to its low profile configurations. The windings consist of either copper spiral turns on single or double-sided printed circuit boards or stamped, flat copper foil windings.

The windings are separated from each other and the core by thin sheets of Kapton or Mylar insulation. When necessary, they are separated by special plastic bobbins to maintain safety specifications. These assemblies are available in a fully encapsulated form to meet certain military specifications.

Advantages of the planar construction include low profile configuration (e.g., high efficiency, lightweight, and uniform construction); high power density (e.g., low manufacturing times, and high frequencies of operation). Some disadvantages include high costs of design and tooling for printed circuit boards and special ferrite cores, inefficient means of terminating windings to the board, and thermal temperature rise of magnetics.

Another approach to planar magnetics is the multilayer circuit board, a more condensed configuration. The primary and secondary windings are stacked vertically on different layers of a circuit board with a thin film of FR4 separating them.

The multilayer circuit board offers all the advantages of the planar configuration and provides an even lower profile transformer. An additional advantage of the multilayer build is that heat is transferred out of the transformer more efficiently so that thermal rise is not as much of a problem compared to single or double layer types.

4.3.2.1 Planar Magnetics and New PWM ICs

With the distributed power architectures (discussed later) being widely adopted by the industry recently, new D/CMOS PWM ICs, combined with integrated PCB transformers, allow simple flyback, and forward converters to operate at frequencies up to 1 MHz, with performance similar to zero switching topologies. An example of such an IC is Si 9114 from Siliconix Inc. (Mohandes 1994).

Planar transformers can be more efficiently used if the subsystem is viewed as an entire unit that includes the power supply and the load. Usually the load is some type of digital circuit and will include a microprocessor and memory, as well as custom designed ASICs. System complexity has so increased that many designers are now using multilayer printed circuit boards where six and even more layers have become common. These systems can now utilize a "free transformer" directly in the PC board as shown in Figure 4–9.

Persson (1994), Magnetics Inc. (1997), Martin (1983), Horgan (1994), and Magnetics Inc. (1995) provide practical details related to magnetic materials and cores used in switchmode power supply systems.

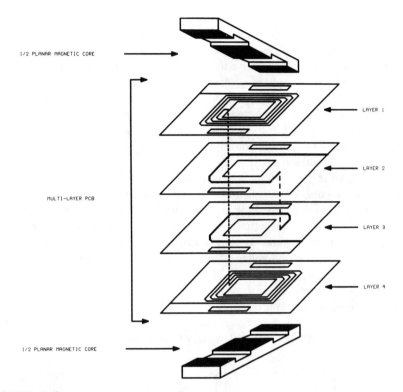

FIGURE 4-9 Integrated PC board planar transformer

4.4 Output Section

In general the output section of any switching power supply is comprised of single or multiple DC voltages, which are derived by direct rectification and filtering of the transformer secondary voltages and in some cases further filtering by series-pass regulators. These outputs are normally low-voltage, direct current, and capable of delivering a certain power level to drive electronic components and circuits. The most common output DC voltages are 5V, 12V, 15V, 24V, or 28V DC, and their power capability may vary from a few watts to thousands of watts.

The most common type of secondary voltages that have to be rectified in a switching power supply are high-frequency square waves. These in turn require special components, such as Schottky or fast recovery rectifiers, low ESR capacitors, and energy storage inductors, in order to produce low noise outputs useful to the majority of electronic components.

4.4.1 Output Rectification and Filtering Schemes

The output rectification and filtering scheme used in a power supply depends on the type of supply topology the designer chooses to use. The conventional flyback converter uses the output scheme shown in Figure 4–10(a). Since the transformer T_1 in the flyback converter also acts as an energy storing inductor, diode D_1, and capacitor C_1 are the only two elements necessary to produce a DC output.

Some practical designs, however, may require the optional insertion of an additional LC filter, shown in Figure 4–10(a) to suppress high-frequency switching spikes. The physical and electrical values of both L_1 and C_2 will be small. For the flyback converter, the rectifier diode D_1 must have a reverse voltage rating of $[1.2V_{in} (N_s/N_p)]$, minimum.

The output section of a forward converter is shown in Figure 4–10(b). Notice the distinct differences in the scheme compared to the flyback. Here, an extra diode D_2 called the flywheel, is added and also inductor L_1 precedes the smoothing capacitor. Diode D_2 provides current to the output during the off period. Therefore, the combination of diodes D_1 and D_2 must be capable of delivering full output current, while their reverse blocking voltage capabilities will be equal to $[1.2V_{in} (N_s/N_p)]$, minimum.

The output section that is shown in Figure 4–10(c) is used for push-pull, half-bridge, and full-bridge converters. Since each of the two diodes D_1 and D_2 provides current to the output for approximately half of the cycle, they share the load current equally. An interesting point is that no flywheel diode is needed, because either

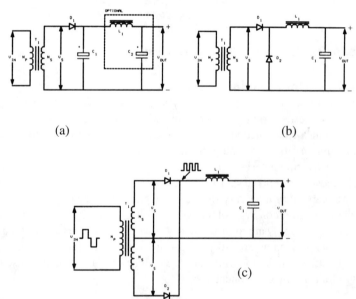

FIGURE 4–10 Output stages (a) Flyback type (b) Forward type (c) Push-pull/half-bridge/full-bridge

diode acts as a flywheel when the other one is turned off. Either diode must have a reverse blocking capability of [$2.4V_{out} (V_{in,max}/V_{in,min})$], minimum.

4.4.2 Magnetic Amplifiers and Secondary Side Regulators

The magnetic-amplifier technology has been around for a long time, but renewed interest has emerged lately, especially with the proliferation of the switching regulator, as an efficient means of regulating the auxiliary outputs of a multiple-output switching power supply. Mag amp output regulators have recently become very popular as a means of regulating more than one output of a switching power supply. They offer extremely precise regulation of each independent output, and are efficient, simple, and very reliable. Magnetic-amplifier usage is universal, that is, they work equally well with any converter topology such as forward, flyback, and push-pull and their derivatives.

Magnetic amplifier or a mag amp is an inductive element wound on a core with relatively square B-H characteristics and used as a control switch. This reactor has two distinct modes of operation: when unsaturated, it acts as an inductance capable of supporting a large voltage with very little or no current flow, and when saturated, the reactor impedance drops to zero, allowing current to flow with zero voltage drop.

Mag amps are particularly well suited for outputs with currents of 1 A to several tens of amps. They are also used at lower currents where tight regulation and efficiency are extremely important.

A mag amp regulates the output of a switched-mode power supply against line and load changes by acting as a fast on/off switch with a high gain. It controls an output by modifying the width of the pulse which appears at the appropriate secondary of the power transformer, before the pulse is "averaged" by the output filter. It does this by delaying the leading edge of the pulse, in the same manner as a series switch which would be open during the first portion of the pulse, and then closed for the balance duration of the pulse.

The advent of cobalt-based amorphous alloy has made possible the design of mag amps that can operate at higher frequencies. Cores made from this alloy exhibit: (i) a high squareness ratio, giving rise to low saturated permeability (ii) low coercive force, indicating a small reset current and (iii) low core loss, resulting in a smaller temperature rise. Although there are alternatives for post output regulation, mag amp regulation is the most efficient, reliable, and low cost design.

Referring to Figure 4–11(a) and waveforms in Figure 4–11(b), assume that at V_1 a waveform varying between $+V_{max}$ and $-V_{max}$ appears. This is from the switching transformer's secondary N_s and it is assumed to be square. At time $t=t_1$, V_1 switches negative. Since the mag amp is saturated, it had been delivering $-V_{max}$ at $t=t_1$ to V_2 (ignoring diode drops).

The control voltage from mag amp control circuit V_c becomes $-V_{cont}$ at t_1 allowing ($V_{max}-V_{cont}$) across mag amp for a T_1 period (t_2-t_1) when V_1 is negative. This negative voltage with the net ($V_{max}-V_{cont}$) value for T_1 period drives the mag amp out of saturation and resets it by an amount equal to ($V_{max}-V_{cont}$)T_1.

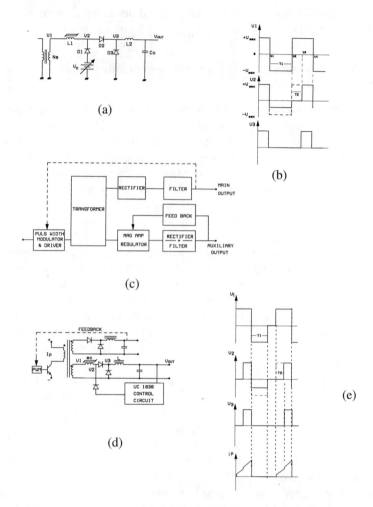

FIGURE 4-11 Magnetic amplifier (a) Basic operation (b) waveforms (c) Two output forward converter with mag amp (d) Mag amp control using UC 1838 (Reproduced with permission of Unitrode Inc., USA) (e) Associated waveforms for two output forward converter

When t=t_2, V_1 switches back to +V_{max}, the mag amp now acts as an inductor and prevents the current from flowing, holding V_2 at zero volts. This condition remains until the voltage across the core (now +V_{max}) back into saturation. The important fact is that this takes the same volt.second value of (V_{max}−V_{cont}) T_1 that was put into core during reset. Therefore:

$$(V_{max} - V_{cont})T_1 = V_{max}(t_3 - t_2) = V_{max} \cdot T_2 \qquad (4.2)$$

This allows the V_o value to be regulated by the pulse duration of the on time. Figure 4–11(c) shows how a mag amp interrelates in a two-output forward converter illustrating the contribution of each output to primary current. Also shown is the use of the UC1838 IC from Unitrode as the mag amp control element in Figure 4–11(d) and 4–11(e) indicates the relevant waveforms.

In addition to regulating, mag amps must often provide current limiting of the individual output. In these cases the saturable reactor must withstand the entire volt-second product of the input waveform. Thus, there are two categories of application: regulation only and shutdown.

Elias (1994) and Mammano and Mullet provide design details of mag amps. Another recent approach (Unitrode Inc. 3/97 and 4/97) for regulating the individual secondary voltages is to use series connected MOSFETs together with secondary side post regulator ICs. Some of these new integrated circuits such as UCC 1583 from Unitrode Inc. are proposed as replacement for mag amp circuits.

4.4.3 Power Rectifiers for Switching Power Supplies

The switching power supply demands that power rectifier diodes must have low forward voltage drop, fast recovery characteristics, and adequate power handling capability. Ordinary PN junction diodes are not suited for switching applications, basically because of their slow recovery and low efficiency. Three types of rectifier diodes are commonly used in switching power supplies:

(a) High-efficiency fast recovery rectifiers
(b) High-efficiency very fast rectifier
(c) Schottky barrier rectifiers

Figure 4–12 shows the typical forward characteristics of these diode types. It can be seen from the graphs that Schottky barrier rectifiers exhibit the smallest forward voltage drop and therefore provide higher efficiencies. Another newer device family is the GaAs power diodes discussed in Chapter 2.

4.4.3.1 Fast and Very Fast Recovery Diodes

Fast and very fast recovery diodes have moderate to high forward voltage drop, ranging from 0.8 to 1.2 V. Because of this and because of their normally high blocking voltage capabilities, these diodes are particularly suited for low-power, auxiliary voltage rectification for outputs above 12 V.

Because most of today's power supplies operate much above 20KHz, the fast and very fast recovery diodes offer reduced reverse recovery time t_{RR} in the nanosecond region for some types. Usually the rule of thumb is to select a fast recovery diode which will have a t_{RR} value at least three times lower than the switching transistor rise time.

These types of diodes also reduce the switching spikes that are associated with the output ripple voltage. Although "soft" recovery diodes tend to be less noisy, their longer t_{RR} and their higher reverse current I_{RM} create much higher switching

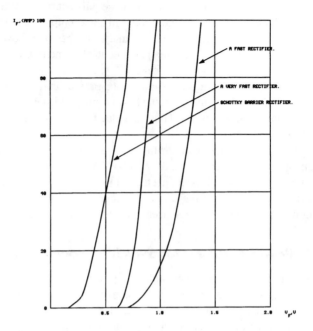

FIGURE 4–12 Typical forward voltage drop characteristics of different rectifier types

losses. Figure 4–13 shows the reverse recovery characteristics of abrupt and soft recovery type diodes.

Fast and very fast switching diodes used in switchmode power supplies as output rectifiers may or may not require heat sinking for their operation, depending upon the maximum working power in the intended application. Normally these diodes have very high junction temperatures, about 175°C, and most manufacturers give specification graphs, which will allow the designer to calculate the maximum output working current versus lead or case temperature.

For example, a new fast recovery diode family with 600V, 15A capabilities termed ultra Fast Recovery Epitaxial Diode (FRED) from International Rectifier exhibits extremely fast reverse recovery times, very low reverse recovery current, unusually "soft" recovery characteristics, and guaranteed avalanche (Khersonsky, Robinson, and Gutierrez 1992).

4.4.3.2 Schottky Barrier Rectifiers

The graph in Figure 4–12 reveals that the Schottky barrier rectifier has an extremely low forward voltage drop of about 0.5V, even at high forward currents. This fact makes the Schottky rectifier particularly effective in low voltage outputs, such as 5V, since in general these outputs deliver high load currents. Moreover, as junction temperature increases in a Schottky, the forward voltage drop becomes even lower. Reverse recovery time in a Schottky rectifier is negligible, because this

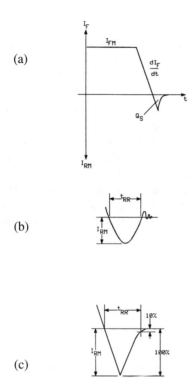

FIGURE 4-13 Behavior of a rectifier as it is switched from forward conduction to reverse at a specific ramp rate $\frac{dI_F}{dt}$ (a) Basic behavior of switching (b) Abrupt recovery type (c) Soft recovery type

device is a majority-carrier semiconductor and therefore there is no minority-carrier storage charge to be removed during switching.

Unfortunately, there are two major drawbacks associated with Schottky barrier rectifiers. First, their reverse blocking capability is low, at present time approximately 100V to 150V. Second, their higher reverse leakage current makes them more susceptible to thermal runaway than other rectifier types. These problems can be averted, however, by providing transient overvoltage protection and by conservative selection of operating junction temperature.

4.4.3.3 Transient Over-Voltage Suppression

Consider the full-wave rectifier, shown in Figure 4–10(c) using Schottky rectifiers D_1 and D_2 in a PWM regulated half-bridge power supply. The voltage V_S across each half of the transformer secondary is $2V_{out}$ minimum; therefore each diode must be capable of blocking $2V_S$ at turn-off, or $4V_{out}$.

Unfortunately, the leakage inductance of the high-frequency transformer and the junction capacitance of the Schottky rectifier form a tuned circuit at turn-off, which introduces transient overvoltage ringing, as shown in Figure 4–14. The amplitude of this ringing may be high enough to exceed the blocking capabilities of the Schottky rectifiers, driving them to destruction during the turn-off period.

The addition of RC snubber networks will suppress this ringing to a safe amplitude, as shown in the lower waveform of Figure 4–14. There are two ways of incorporating RC snubbers at the output of a power supply to protect the Schottky rectifiers. For high current outputs the snubbers are placed across each rectifier as shown in Figure 4–15(a) while for low current outputs a single RC snubber across the transformer secondary winding, as shown in Figure 4–15(b) may be adequate.

Another solution is to place a zener diode, as shown in Figure 4–15(c) to clamp the excessive voltage overshoot to safe levels. Although this method works well, the slow recovery of a zener may induce noise spikes at the power supply output which may not be desirable for low noise applications. For calculation details of RC components of snubber circuits Chryssis (1989) is suggested.

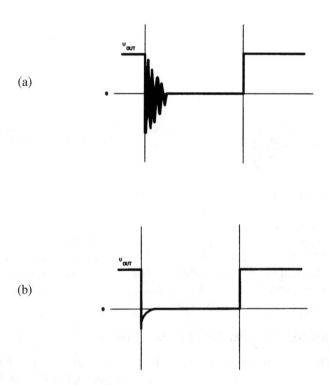

FIGURE 4–14 Ringing during Schottky rectifier turn-off (a) With no snubber (b) With snubber

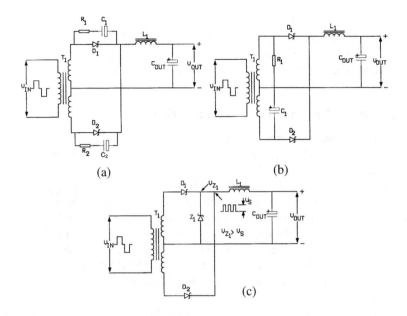

FIGURE 4-15 Means of protecting output Schottky rectifiers during turn-off (a) Snubbers placed across each rectifier (b) A single RC snubber placed across the transformer (c) Using a zener diode

4.4.3.4 Synchronous Rectifiers

As digital integrated circuit manufacturers are working toward implementing more electrical functions and circuits on a single silicon chip, the need for lower bias voltages becomes evident. For example, there are new families of logic circuits which require 3.3V supply.

These new voltage standards have forced design engineers of switching power supplies to explore the possibility of using devices other than junction diodes for rectification, in order to reduce the power losses encountered at the output section. Since the power dissipation is directly proportional to the diode's forward voltage drop, even the best Schottky rectifiers with a 0.4V forward voltage drop may account for up to 20 percent of the total input power loss.

In recent years the power semiconductors such as bipolar power transistors and power MOSFETS have provided an alternative in power rectification and these devices working as synchronous rectifiers have started to appear in switching power supply designs. Use of ultra-low $r_{DS,(on)}$ MOSFETs as synchronous rectifiers may provide the best solution for overcoming the single most important barrier to higher power supply efficiency.

Low $r_{DS,(on)}$ MOSFETs employ VLSI processing and design techniques that allow manufacture of low-voltage MOSFETs with more parallel cells packed into given die area. The result is a reduced channel resistance, which is the dominant component of $r_{DS,(on)}$ for devices with breakdown voltages under 60 V. For example, a process developed by Siliconix Inc., USA enables 2.5 million MOSFET cells per square inch of die with good manufacturing yields. This process allows manufacturing economical MOSFETs with 18 mΩ maximum on resistance in a TO-220 package (Mohandes 1991).

4.4.3.4.1 Practical Circuit Configurations

Figure 4–16 shows a practical implementation of a center-tapped full-wave output section of a switching power supply utilizing power MOSFETs as synchronous rectifiers.

MOSFETs Q_1 and Q_2 are selected to have minimum possible $R_{DS,(on)}$ at maximum output current, in order to increase efficiency. Transformer windings N_3 and N_4 are used to turn on the MOSFETs at opposite half cycles of the input waveform. It is beneficial to overdrive the gate-to-source voltage in order to minimize $R_{DS,(on)}$.

Careful selection of resistors R_1 through R_4 will also minimize the switching times. Diodes D_1 and D_2 are parasitic diodes of the MOSFETs, and act as freewheeling diodes during the input wave-form dead time t_d to provide current path for inductor L_1.

Figure 4–17 shows the implementation of the MOSFET as synchronous rectifier in a single-ended forward converter. In this configuration transistor Q_1 turns on by the gate-to-source voltage developed by winding N_1, on the positive cycle of the switching waveform. When the input voltage swings negative, MOSFET Q_1 turns off and output current is supplied to the load through MOSFET Q_2, which now turns on through coupled winding N_3 using energy stored in the output inductor L.

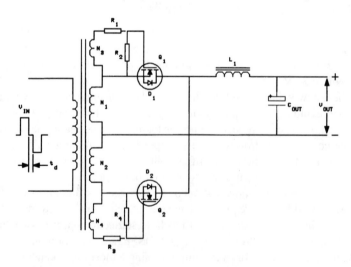

FIGURE 4–16 Full-wave output rectification of a switching power supply using MOSFETS as synchronous rectifiers

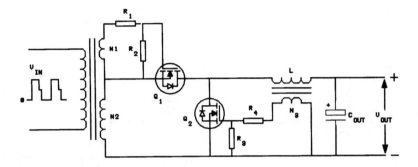

FIGURE 4-17 Circuit implementation of a forward converter output section using power MOSFETS as synchronous rectifiers.

The criteria for choosing the proper MOSFET parameters for use in this circuit are the same as in the bridge configuration output circuit.

Synchronous rectifiers implemented with special bipolar transistors offer all the advantages of the MOSFET synchronous rectifiers up to frequencies of 200 KHz (above this frequency the MOSFET dominates), coupled with lower prices and improve temperature coefficients compared to power MOSFETs. For details of practical implementation of bipolar synchronous rectifiers Chryssis (1989) is suggested.

4.4.4 Inductors and Capacitors in the Output Section

4.4.4.1 Inductors and Capacitors: A Comparison

Inductors and capacitors are used in the output section as energy storage elements. Table 4–5 is a comparison of some capacitors and inductors to store the energy of 0.1 joule. The data is based on information as per Persson (1994) prices. Data may vary depending on selections. Using this table as a guideline one could easily see that inductors are bulky, costly, and heavy. However for the reasons given below inductors and capacitors can not be just interchanged.

    a. Capacitors behave like low impedance voltage sources
    b. Inductors behave like high impedance current sources

Most switching power supply designs use an inductor as part of their output filtering configuration. The purpose of this inductor is two-fold. First, it stores energy during the off or "notch" periods in order to keep the output current flowing continuously to the load. Second, it aids to smooth out and average the output voltage ripple to acceptable levels.

TABLE 4–5 Comparison of Capacitors and Inductors as Energy Storage Elements

Component	Approximate Volume mm^3 (in^3)	Approximate Weight grams (lbs)	Approximate Cost (US$)
300µF, 25V Electrolytic Capacitor	1000 (.06)	1.3 (.003)	0.20
1.2µF, 400V film capacitor	5000 (0.3)	9.0 (0.02)	1.50
1 mH, 14Amp Powdered Iron Inductor (T4000D)	311x10³ (19)	1.36k (3.0)	12.00
200µH, 30A gapped ferrite core inductor (2xEC-70)	590x10³ (36)	0.91k (2.0)	25.00

4.4.4.2 Output Filter Capacitors

Ripple filter capacitors are major factors in reducing the size and weight of power conversion systems. Typical capacitors for this application include tantalums, electrolytics, multilayer ceramic (MLC), and the Multi-layer Polymer (MLP) capacitors.

Two of the most important high frequency SMPS capacitor characteristics are equivalent series resistance, ESR, and equivalent series inductance, ESI. These two parameters affect the output ripple voltage, output noise, and supply efficiency. When SMPS technology was in its infancy, no single output capacitor could meet low ESR requirements, so several capacitors were used in parallel to meet ripple current specifications.

Output filters in high frequency SMPS units require extremely low ESR and ESL capacitors. These requirements are beyond the practical limits of electrolytic capacitors, both aluminum and tantalum. However, MLC and MLP capacitors meet these requirements.

At high switching frequencies, filter inductance or ESL dominates output ripple voltage. MLC output filter capacitors have ESR and ESL values well below the minimum values needed resulting in lower noise levels at high frequencies, thus eliminating the need for parallel output filter capacitors. Currently, one of the driving forces for size reduction in higher frequency supplies is use of surface mount technology (SMT). MLCs are compatible with SMT reflow and assembly techniques.

Much newer process techniques used in the MLP capacitors, offer excellent electrical stability under AC and DC loads and are not subject to the cracking, shorting, or temperature coefficient mismatches inherent in MLC types (Clelland and Price 1994). The MLP capacitor is fully surface mount compatible and is best described as a construction hybrid between MLC and stacked, plastic film capacitors.

The present offering of non-polar, highly stable MLP capacitors covers the range from 0.0047µF through 20µF with voltages from 25 to 500 VDC. Today, this technology is leading in C * V density and is growing most rapidly in input voltage filtering from 48 to 400V, and output filtering at 24 to 48V. Many of the products are available in surface mount styles as either lead-framed construction or as true "chip" capacitors. Some of the commercially available MLP capacitor ranges are Angstor, Surfilm, and Capstick. Table 4–6 indicates the applications, and ranges applicable in these MLP capacitors.

TABLE 4–6 MLP Capacitors and Applications (Source: Clelland and Price 1994)

Type	Voltage Range (Volts)	Capacitor Value (μF)	Applications
Angstor	50-450	.0047-3.3	General
Surfilm	25-50	.01-2.2	Standard EIA Chip sizes
Capstick	50-100	4.0-20	Telecom products
Capstick	400DC/250 AC	.33-1.0	Off-line systems/PFC front ends

For further details on MLC and MLP capacitors references Clelland (1992), Lienemann (1992), Clelland and Price (1993, 1994, 1997), and Cygan (1997) are suggested.

4.5 Ancillary, Supervisory, and Peripheral Circuits

In general, switching power supplies are closed-loop systems, a necessity for good regulation, ripple reduction, and system stability. Other than the basic building blocks, which have been described in the previous sections, there are a number of peripheral and ancillary circuits that enhance the performance and reliability of the power supply.

Components such as opto-isolators are extensively used in the flyback or feed-forward converter to offer the necessary input-to-output isolation and still maintain good signal transfer information. Other circuits, such as soft-start, over-current, and overvoltage protection circuits, are used to guard the power supply against failures due to external stresses. The details of such circuits and components are beyond the scope of a chapter of this kind.

4.6 Power Supplies for Computers

In today's electronics marketplace, power supply vendors face a tremendous market for the various power supplies required by computer systems. For the most popular models of portable, desktop, and minicomputer systems, manufacturers demand regulated power supplies with direct AC line operation with output capacities of approximately 25 to 500W; multiple low DC voltage of 5V, 12V, –12V, and

–5V derived from the same ferrite transformer with multiple secondary windings; individual regulation using three terminal regulators; monolithic primary side controller chips to switch the DC voltage across the ferrite transformer to provide the necessary isolation from the low-voltage output; RFI and EMI filters installed at the incoming end of the AC supply to prevent high-frequency pollution generated by the SMPS; and the various supervisory and control circuits required for reliable SMPS operation.

In most computer power supplies (Figure 4–18) multiple outputs are under the control of a single SMPS controller chip. Depending on the power output, various auxiliary circuits may be powered from independent SMPS modules or linear blocks, especially in high capability minicomputer power supplies.

The latest microprocessors to emerge from Intel, Motorola, and others are forcing fundamental changes in the power supplies for desktop and portable computers. But not only do the μPs demand lower and more precise supply voltages, but their main clocks also exhibit start/stop operation that causes ultra-fast load transients. As a result, the relatively simple 5V/12V supply has been transformed into a system with five or more outputs, featuring unprecedented accuracy and 50A/μs load-current slew rates.

These characteristics present a problem: it appears that the classic, centralized power-supply architecture cannot provide the accuracy and transient response needed by coming generations of computer systems. The more effective approach will be a distributed architecture in which local, highly efficient DC-DC converters are located on the mother board next to the CPU.

Product designers could expect power-supply manufacturers to respond with smaller, higher-frequency ICs and modules that feature improved dynamic response and better synchronous rectifiers. The PC's off-line power supply will not disappear. It will remain to generate the main bus for small DC-DC converters on the motherboard.

4.7 Modular SMPS Units for Various Industrial Systems

With a heavy demand from industrial system designers for different series of low-power (less than 100W) and low-voltage power supply blocks, many manufacturers have introduced various kinds of DC/DC converter modules for various fixed output voltages such as 5V, ±12V, ±15V, 24V, 28V, and ±48V for off-the-shelf selection. The same manufacturers market similar AC/DC power supply modules, among them the manufacturers such as Vicor, Lambda, Kepco, and Datel.

Another new series of devices coming to the market are integrated switching regulators. Units such as those by Power Trends (Figure 4–19) are similar to the popular three-terminal linear voltage regulators.

One recent advancement was to move power MOSFET from its discrete package and heat sink on to the same silicon die as its controller, building a power IC. With this new trend, designs could minimize the total component count of an off-the-line SMPS. Power switches and the controller packaged in one IC could be used with few other external magnetic components and the feedback loop requirements.

Off-the-Line Switchmode Power Supplies   131

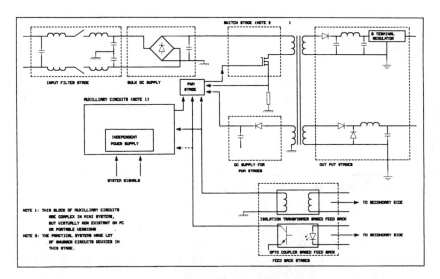

FIGURE 4-18   A representative computer power supply

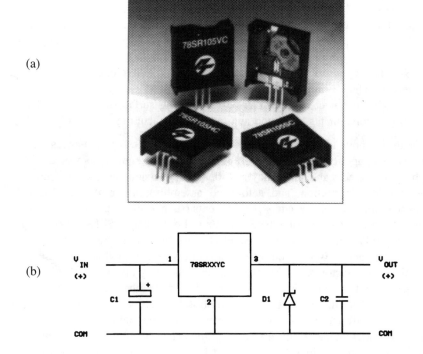

FIGURE 4-19   Integrated switching regulator modules (Reproduced by permission of Power Trends Inc., USA.) (a) 3 terminal modules (b) A typical application

Classic examples of these are the TOPswitch series from Power Integration and VIPer 100 series from SGS Thompson. These switchers offer power outputs in the 10 to 150W range, which is sufficient for modern desktop PCs and similar applications.

TOP switches are available for universal wide-input-range applications such as 85 to 260V or 110/115/230V AC inputs. Similarly the VIP (Vertical Integrated Power) family, which is a 600V rated component, could operate from a rectified 85/115V AC line. These components, operating at lower frequencies (100 to 200 kHz), could virtually eliminate EMI/RFI problems too. For details Goodenough (1997) is suggested.

4.8 Future Trends in SMPS

In this decade, power conversion designs will lead towards decentralized distributed systems. Technically, this will create a situation in which power-hungry systems will contain one central DC power supply distributed to all subassemblies of PCBs with local SMPS units powering the circuitry. Four key factors listed below are dictating the move toward distributed power.

- Central power supplies no longer support the advanced semiconductors that operate at lower supply voltages, run at greater speeds and demand more power.
- Custom power supplies require a long development cycle and are costly. Distributed power with parallel modules are more adaptable to design standards.
- Redundancy with parallel modules permits improved reliability, easy maintenance and on-line repair.
- Distributed power supplies can be made using surface-mount technology (SMT) components to support automated manufacturing.

With the move towards distributed power, several trends and changes affect the design and application of power supplies, including: greater power density (switching frequency and efficiency); resonant conversion techniques (surface mount, hybrid construction, and automated assembly); low-profile magnetics; changes in extra-low voltage safety requirements (such as the International Electro Technical Commission changing the Safety Extra Low Voltage (SELV) limits from DC 42.4V to 60V); and the development of DC uninterruptible power supplies.

An important technical requirement of future switched-mode power supplies will be the power factor correction that will be mandatory above a certain power consumption limit, such as 700 VA. This is because of the inherent characteristic of switchmodes, where the input current waveforms from AC line will be non-sinusoidal (and pulsed) and the peak of the current waveform and the peak AC input voltage will not coincide. To meet such requirements, several companies produce power factor controller chips such as Micro Linear's ML4812 and ML4813 that help to achieve a nearly unity power factor SMPS. Kularatna (1992) and Maxim Inc. (1995) provide a detailed account of future trends.

4.9 Field Trouble-Shooting of Computer Systems Power Supplies

In environments where power quality on the commercial AC lines is not up to standards, such as in some developing countries, a large percentage (perhaps 25 percent) of hardware faults on mini, micro, and portable computers occur in the power supplies. To begin with a fault trace procedure, the input stage that carries the AC power into the bulk DC stage (via an RFI/EMI filter) should be checked for the availability of AC line voltage after the filter stage (following a careful visual inspection for physically damaged power components).

If this stage is found to be fault-free, the next step is to check the bulk DC stage where reasonably high DC voltages (up to 350V in cases of AC 220V, 50Hz inputs) are present. An oscilloscope should be used to make sure that an excessive ripple component is not present, superimposed on the bulk DC.

Once troubleshooting has confirmed the availability of bulk DC, the next step is to check the PWM controllers, using circuit diagrams to identify the topology of the switching stage and the type of power semiconductors used for switching. In most computer power supplies a large number of snubbers and clamping circuits (based on RC components or zener diodes) etc. are used around the drive stages for reducing the transients below the maximum ratings of power semiconductors.

It is always safe and saves time to check the correct operation of zener diodes or other semiconductor devices around the drive circuits and the main power elements before using the scope to check the drive pulses. In many cases of power transistor failures, several other semiconductors around the power transistors also may be faulty.

If the main power semiconductors are operational and drive pulses are available, the voltage- or current-related feedback paths should be checked. In cases where the pulses driving the main switch elements are not observed, the feedback operation of the SMPS controller chips should be checked. In most cases, the output voltage feedback signal is derived from the output power rail that drives the maximum VA rating. In a personal computer power supply, most power is drawn from the 5V rail.

In cases where none of the output voltages are present, it might be worth suspecting the primary side of the SMPS rather than placing priority on the individual regulators and other associated circuitry on the output power rails. In the case of complex minicomputer power supplies, there may be many control inputs coming from the auxiliary circuits which act as safety measures to stop or regulate the overall system. In such cases, it may be necessary to identify those areas and signals that may be inhibiting the final drives.

Once the primary side oscillation is achieved, most of the basic trouble-shooting is achieved. When one or two output rails have their specified voltages, it is a matter of detailing the secondary side for individual rail faults. In practice, a dummy load may need be connected to the 5V output in case intermittent oscillations or unregulated outputs are observed.

References

1. Chryssi, George C. *High Frequency Switching Power Supplies*, 2nd Edition. McGraw-Hill, 1989.
2. Magnetics Inc. "EMI/RFI Filter-Common Mode Filter." Technical Bulletin FC-S2 4C, 1995.
3. Billings, Keith. "Switchmode Power Supply Handbook." McGraw-Hill, 1989.
4. Brown, Marty. *Practical Switching Power Supply Design*. Academic Press Inc., 1990.
5. Magnetics Inc. "Magnetic Cores for Switching Power Supplies." 1995.
6. Visser, Eelco G. and Alex Shpilman. "New Power Ferrite Operates from 1 to 3 MHz."*PCIM*, April 1991, pp 17–23.
7. Bates, Gary. "New Transformer Technologies Improve Switch Mode Power Supplies." *PCIM*, July 1992, pp 28-31.
8. Mohandes, Bijan E. "Integrated PC Board Transformers Improve High Frequency PWM Converter Performance." *PCIM*, July 1994, pp 8–17.
9. Magnetics Inc. "Magnetic Cores for Switching Power Supplies." Publication-PS-02 2E,1997.
10. Martin, A. W. "Magnetic Cores for Switching Power Supplies." Magnetics Inc., App note, 1983.
11. Horgan, M. W. "Comparison of Magnetic Materials for Flyback Transformers." *PCIM*, July 1994, pp 18-24.
12. Magnetics Inc. "Inductor design in switching regulators." Technical Bulletin SR-1A, 1995.
13. Elias, J. S. "Design of High Frequency Mag Amp Output Regulators Using Metglas Amorphous Alloy 2714A." *EPE Journal*, Vol 4,No 4, December 1994, pp 19–24.
14. Persson, Eric. "The ABCs of Magnetics for Digital Engineers." HFPC conference proceedings, April, 1994, pp 88–106.
15. Mammano, R. A. and C. F. Mullet. "Using an integrated controller in the design of mag-amp output regulators." Unitrode application note-U-109.
16. Khersonsky, Y., M. Robinson, and D. Gutierrez. "New Fast Recovery Diode Technology Cuts Circuit Losses, Improves Reliability." *PCIM*, May 1992, pp 16–25.
17. Mohandes, Bijan. "MOSFET Synchronous Rectifiers achieve 90% efficiency." *PCIM*, June 1991, pp 10–13.
18. Clelland, Ian. "Metalized Polyester Film Capacitor Fills High Frequency Switcher Needs." *PCIM*, June 1992, pp 21–30.
19. Lienemann, Dennis. "ESR and ESL Affect High Frequency SMPS Capacitor Selection." *PCIM*, June 1992, pp 8–14.
20. Clelland, I. W. and R. A. Price. "Precision Film Capacitors for High Frequency Circuits." Proceedings of HFPC, May 1993, pp 1–10
21. Clelland, I. W. and R. A. Price. "Multilayer Polymer (MLP) Capacitors Provide Low ESR and are Stable over Wide Temperature and Voltage Ranges." Proceedings of the European Capacitor and Resistor Technology Symposium, October 1994.
22. Carsten, Bruce. "Optimizing Output Filters Using Multilayer Polymer Capacitors in High Power Density Low Voltage Converters." Technical Bulletin 5.95, ITW Paktron, USA.
23. Clelland, I. W. and R. A. Price. "Ultralow ESR Multilayer Polymer Capacitors Provide Stability and Reliability." *PCIM*, June 1997, pp 30–35.
24. Cygan, P. S. "Multilayer Ceramic Capacitors Meet HF SMPS Requirements." *PCIM*, March 1997, pp 8-19.
25. Kularatna, A.D.V.N. "Switchedmode Power Supplies Offer Design Flexibility." *AEU*, 1992, pp 83–91.
26. Davis, Sam. "SMT Passive Components Fit Power Electronics Applications." *PCIM*, June 1993, pp 20-28.
27. Kerridge, Brian. "High Power Modular Switching Power Supplies—Custom Configured Supplies Promote Design Flexibility." *EDN*, 7 May 1992, pp 79–86.
28. Ormond, Tom. "Distributed Power Schemes Put power Where You Need It." *EDN*, 6 July 1992, pp 158–164.

29. Ormond, Tom. "External Power Supplies Enhance System Design Options." *EDN*, 4 February 1993, pp 39–42.
30. Davis, Sam. "High Frequency Power Conversion in the Next Decade." *PCIM*, April 1989, pp 18–22.
31. Maxim Inc. "Power Supplies for Pentium, Power PC, and Beyond." *Maxim Engineering Journal*, 1995 pp 3–7.
32. De Palama, J. F. and R. Deshayes. "High Power Semiconductor Protection Requires the Appropriate Fuses." *PCIM*, October 1996, pp 58–65.
33. Goodenough, F. "Switch-Mode ICs Minimize Off-Line Power Supply Size." *Electronic Design*, April 1, 1997, pp 53–62.
34. Unitrode Inc. "Switchmode Secondary Side Post Regulator—UCC 1583/2583/3583 data sheets (preliminary)." 3/97.
35. Unitrode Inc. "Switchmode Side Synchronous Post Regulator—UCC 1584/2584/3584 data sheets (preliminary)." 4/97.

CHAPTER **5**

Rechargeable Batteries and Their Management

5.1 Introduction

The insatiable demand for smaller lightweight portable electronic equipment has dramatically increased the need for research on rechargeable (or secondary) battery chemistries. In addition to achieving improved performance on Lead Acid and Nickel Cadmium (NiCd) batteries, during the last two decades many new chemistries such as Nickel Metal Hydride (NIMH), Lithium Ion (Li-Ion), Rechargeable Alkaline, Silver-Zinc, Zinc-Air, Lithium Polymer, etc. have been introduced.

Higher Energy density, superior cycle life, environmentally friendliness and safe operation are among the general design targets of battery manufacturers. To complement these developments many semiconductor manufacturers have introduced new integrated circuit families to achieve the best charge/discharge performance and longest possible lifetime from battery packs.

This chapter describes the characteristics of battery families such as Sealed Lead Acid, NiCd, NIMH, Li-Ion, and Rechargeable Alkaline together with modern techniques used in battery management ICs, without elaborating on the battery chemistries. A brief introduction to Zn-air batteries is also included.

5.2 Battery Terminology

5.2.1 Capacity

Battery or cell capacity means an integral of current over a defined period of time.

$$\text{Capacity} = \int_0^t i \, dt \qquad (5.1)$$

This equation applies to either charge or discharge, i.e., capacity added or capacity removed from a battery or cell. The capacity of a battery or cell is measured in milliamperes-hours (mAh) or ampere-hours (Ah).

Although the basic definition is simple, many different forms of capacity are used in the battery industry. The distinctions between them reflect differences in the conditions under which the capacity is measured.

5.2.1.1 Standard Capacity

Standard Capacity measures the total capacity that a relatively new, but stabilized production cell or battery can store and discharge under a defined standard set of application conditions. It assumes that the cell or battery is fully formed, that it is charged at standard temperature at the specification rate, and that it is discharged at the same standard temperature at a specified standard discharge rate to a standard end of discharge voltage (EODV). The standard EODV is itself subject to variation depending on discharge rate as discussed.

5.2.1.2 Actual Capacity

When any of the application conditions differ from standard, the capacity of the cell or battery change. A new term, actual capacity, is used for all nonstandard conditions that alter the amount of capacity which the fully charged new cell or battery is capable of delivering when fully discharged to a standard EODV. Examples of such situations might include subjecting the cell or battery to a cold discharge or a high-rate discharge.

5.2.1.3 Available Capacity

That portion of actual capacity, which can be delivered by the fully charged new cell or battery to some nonstandard end-of-discharge voltage is called available capacity. Thus, if the standard EODV is 1.6 volts per cell, the available capacity to an EODV of 1.8 volts per cell would be less than the actual capacity.

5.2.1.4 Rated Capacity

Rated capacity is defined as the minimum expected capacity when a new, but fully formed, cell measured under standard conditions. This is the basis for C rate and depends on the standard conditions used which may vary depending on the manufacturers and the battery types.

5.2.1.5 Retained Capacity

If a battery is stored for a period of time following a full charge some of its charge will dissipate. The capacity which remains that can be discharged is called retained capacity.

5.2.2 C Rate

The C rate is defined as the rate in amperes or milliamperes numerically equal to the capacity rating of the cell given in ampere-hours or milliampere-hours. For example, a cell with a 1.2 ampere-hour capacity has a C rate of 1.2 amperes. The C concept simplifies the discussion of charging for a broad range of cell sizes since the cells' responses to charging are similar if the C rate is the same. Normally a 4Ah cell will respond to a 0.4 amp (0.1C) charge rate in the same manner that a 1.4 Ah cell will respond to a 0.14 amp (also 0.1C) charge rate.

The rate at which current is drawn from a battery affects the amount of energy which can be obtained. At low discharge rates the actual capacity of a battery is greater than at high discharge rates. This relationship is shown in Figure 5–1.

5.2.3 Energy Density

Energy density of a cell is its energy divided by its weight or volume. When weight is used it is called the gravimetric energy density, and volumetric energy density when the volume is used.

5.2.4 Cycle Life

Cycle life is a measure of a battery's ability to withstand repetitive deep discharging and recharging using the manufacturer's cyclic charging recommendations and still provide minimum required capacity for the application. Cyclic discharge testing can be done at any of various rates and depths of discharge to simulate conditions in the application. It must be recognized, however, that cycle life has an inverse logarithmic relationship to depth of discharge.

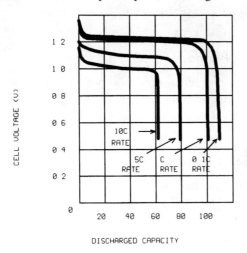

FIGURE 5–1   Capacity vs. discharge rate of a typical cell

5.2.5 Cyclic Energy Density

For purposes of comparison, a better measure of rechargeable battery characteristics is a composite characteristic which considers energy density over the service life of the battery. A composite characteristic, Cyclic Energy Density, is defined as the product of energy density and cycle life at that energy density and has the dimensional units, watt-hour-cycles/kilogram (gravimetric) or watt-hour-cycles/liter (volumetric).

5.2.6 Self-Discharge Rate

Self-discharge rate is a measure of how long a battery can be stored and still provide minimum required capacity and be recharged to rated capacity. It is commonly measured by placing batteries on shelf stand at room (or elevated) temperature and monitoring open circuit voltage over time. Samples are discharged at periodic intervals to determine remaining capacity and recharged to determine rechargeability.

5.2.7 Charge Acceptance

Charge acceptance is the willingness of a battery or cell to accept charge. This is affected by cell temperature, charge rate, and the state of charge.

5.2.8 Depth of Discharge

Depth of discharge is the capacity removed from a battery divided by its actual capacity, expressed as a percentage.

5.2.9 Voltage Plateau

Voltage plateau is the protracted period of very slowly declining voltage that extends from the initial voltage drop at the start of a discharge to the knee of the discharge curve (see Figure 5–2.)

5.2.10 Midpoint Voltage

Midpoint voltage is the battery voltage when 50 percent of the actual capacity has been delivered (see Figure 5–2.)

5.2.11 Overcharge

Overcharge is defined as continued charging of a cell after it has become fully charged. When a cell is not yet fully charged, the electrical energy of the charge current is converted to chemical energy in the cell by the charging reactions. But, when all of the available active material has been converted into the charged state, the energy available in the charging current goes to produce gases from the cell or to activate other nonuseful chemical reactions.

5.3 Battery Technologies: An Overview

Many rechargeable chemistries are available for use in electronic systems. Today, NiCd, NIMH, and Sealed Lead Acid comprise the bulk of the shipments, with Lithium-Ion making headway into portable systems. The choice of a particular battery technology for a given system is typically limited by size, weight, cycle life, and cost. Comparison of basic characteristics of five major chemistries are depicted in Table 5–1.

NiCd batteries presently power most rechargeable consumer appliances. It is a mature, well-understood technology. However, cadmium is coming under increasing regulatory scrutiny (including mandatory recycling in some jurisdictions), and the maturity of NiCd technology also means most of the capacity and life cycle improvements have already been made.

NIMH offers incremental improvements in energy density both by weight and volume over NiCd. Li-Ion is better still, offering over twice the watts per liter and per kilogram of NiCd batteries. As always, this higher performance comes at a higher price. NIMH and Li-Ion and are increasing in popularity as upgrade options or in applications that support a higher price/performance point.

The advantages of NiMH and Li-Ion chemistries, however, also come at the cost of greater electrical fragility. Li-Ion particularly is more easily and extensively damaged by less than optimal battery management so much so that fail-safe circuits to disconnect the cells from the load under over-current or over-temperature conditions are usually built into the battery pack.

Rechargeable alkaline batteries mimic the form and replace the function of disposable household batteries. While initially more expensive, they cost less over their lifetime than their equivalent in disposable batteries. They are the least expensive of the rechargeable chemistries for low current applications and have the lowest self-discharge rate. However, they have the shortest cycle life in deep discharge applications.

TABLE 5-1 Battery Chemistry Characteristics

Parameter	Units/ conditions	Sealed Lead Acid	NiCd	NiMH	Li-Ion	Rechargeable Alkaline
Cell Voltage	Volts	2.0	1.2	1.2	3.6	1.5
Relative cost	NiCd= 1	0.6	1	1.6	2	0.5
Self Discharge	%/month	2%-4%	15%-30%	18%-20%	6%-10%	0.3 %
Cycle Life	cycles to reach 80% of rated capacity	500-2000	500-1000	500-800	1000-1200	<25
Overcharge tolerance		High	Med	Low	Very Low	Med
Energy by volume	watt hour/liter	70-110	100-120	135-180	230	220
Energy by weight	watt hour/kg	30-45	45-50	55-65	90	80

Lead-acid batteries are most familiar in automobiles because they are the most economical chemistry for delivering large currents. Lead acid also has a long trickle life and therefore serves well for classic "floating" applications. While flooded lead acid technology is popular for automobile and similar applications, sealed lead acid batteries serve the electronic engineering environments. On the downside, lead-acid has the least capacity by volume and weight.

Table 5–1 lists the major advantages and disadvantages of the five chemistries. Chemistry selection involves tradeoffs driven by the technical requirements and economics of the application.

5.4 Lead-Acid Batteries

5.4.1 Flooded Lead-Acid Batteries

The flooded lead-acid battery today uses basically the design developed by Faure in 1881. It consists of a container with multiple plates immersed in a pool of dilute sulfuric acid. Recombination is minimal so water is consumed through the battery life and the batteries can emit corrosive and explosive gases when experiencing overcharge.

So-called "maintenance-free" forms of flooded batteries provide excess electrolyte to accommodate water loss through a normal life cycle. Most industrial applications for flooded batteries are found in motive power, engine starting, and large system power backup. Today, other forms of battery have largely supplanted flooded batteries in small and medium capacity applications, but in larger sizes flooded lead-acid batteries continue to dominate.

By far, the biggest application for flooded batteries is starting, lighting, and ignition (SLI) service on automobiles and trucks. Large flooded lead-acid batteries also provide motive power for equipment ranging from forklifts to submarines and provide emergency power backup for many electrical applications, most notably the telecommunications network.

5.4.2 Sealed Lead-Acid Batteries

Sealed lead-acid batteries first appeared in commercial use in early 70s. Although the governing reactions of the sealed cell are the same as other forms of lead-acid batteries, the key difference is the recombination process that occurs in the sealed cell as it reaches full charge. In conventional flooded lead-acid systems, the excess energy from overcharge goes into electrolysis of water in the electrolyte with the resulting gases being vented.

This occurs because the excess electrolyte prevents the gases from diffusing to the opposite plate and possibly recombining. Thus, electrolyte is lost on overcharge with the resulting need for replenishment. The sealed-lead cell, like the sealed nickel-cadmium, uses recombination to reduce or eliminate this electrolyte loss.

Sealed-lead acid batteries for electronics applications are somewhat different from the type commonly found in the automobile. There are two types of sealed lead-acid batteries: the original gelled electrolyte and retained (or absorbed) system. The gelled electrolyte system is obtained by blending silica gel with an electrolyte, causing it to set up in gelatin form. The retained system employs a fine glass fibre separator to absorb and retain liquid electrolyte. Sometimes the retained system is named Absorbed Glass Mat (AGM). AGM is also known in the industry as "starved design."

Starved refers to the absorption limits of the glass separator creating a limitation to the AGM design relating to diffusion properties of the separator. In certain cases the AGM battery must be racked and trayed in a specific position for optimum performance. Both these types, gelled and AGM, are called Valve Regulated Lead Acid (VRLA) systems. Today, sealed-lead cells are operating effectively in many markets previously closed to lead-acid batteries. For a detailed account of Lead-Acid cells references, Energy Products Inc. (1992), Hirai (1990), and Moore (1993) are suggested.

Meanwhile, some manufacturers have introduced special versions of sealed lead-acid batteries with higher volumetric energy density. For example, Portable Energy Products, Inc. (USA) has introduced its Thinline™ Series with comparatively higher energy density (Moneypenny and Wehmeyer (1994)).

5.4.2.1 Discharge Performance of Sealed-Lead Acid Cells

The general shape of the discharge curve, voltage as a function of capacity (if the current is uniform) is shown in Figure 5–2. The discharge voltage of the starved-electrolyte sealed-lead acid battery typically remains relatively constant until most of its capacity is discharged. It then drops off sharply.

144 POWER ELECTRONICS DESIGN HANDBOOK

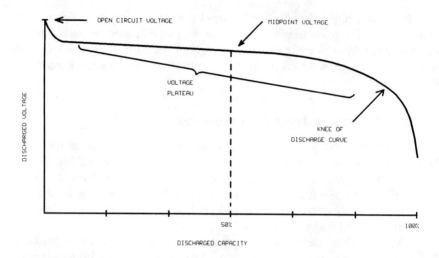

FIGURE 5-2 Nominal discharge performance for sealed-lead acid cells

The flatness and the length of the voltage plateau relative to the length of the discharge are major features of sealed-lead cells and batteries. The point at which the voltage leaves the plateau and begins to decline rapidly is often identified as the knee of the curve.

Starved-electrolyte sealed-lead acid batteries may be discharged over a wide range of temperatures. They maintain adequate performance in cold environments and may produce actual capacities higher than their standard capacity when used in hot environments. Figure 5-3 indicates the relationships between capacity and cell temperature. Actual capacity is expressed as a percentage of rated capacity as measured at 23°C.

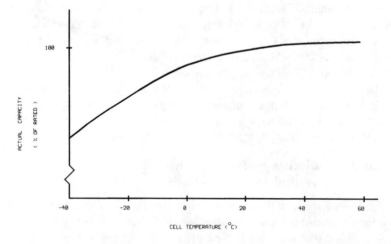

FIGURE 5-3 Typical discharge capacity as a function of cell temperature

5.5.3 Charge Characteristics

Nickel-cadmium batteries are charged by applying a current of proper polarity to the terminals of the battery. The charging current can be pure direct current (DC) or it may contain a significant ripple component such as half-wave or full-wave rectified current.

This section on charging sealed nickel-cadmium batteries refers to charging rates as multiples (or fractions) of the C rate. These C rate charging currents can also be categorized into descriptive terms, such as standard-charge, quick-charge, fast-charge, or trickle-charge as shown in Table 5–2.

When a nickel-cadmium battery is charged, not all of the energy input is converting the active material to a usable (chargeable) form. Charge energy also goes to converting active material into a unusable form, generating gas, or is lost in parasitic side reactions.

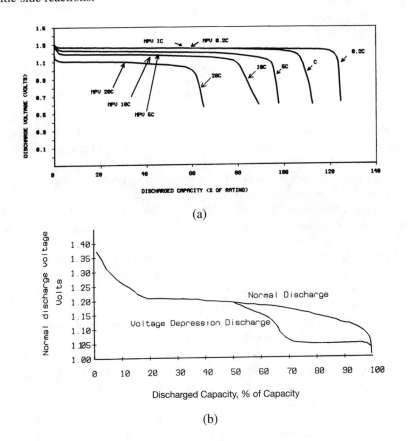

FIGURE 5–7 Discharge curves for NiCd cells (a) Typical curves at 23°C (b) Voltage depression effect

TABLE 5-2 Definition of Rates for Charging NiCd Cell

Method of Charging	Charge Rate		Recharge Time (Hours)	Charge Control
	Multiples of C-Rate	Fractions of C-Rate		
Standard	0.05	1/20	36-48	Not Required
	0.1	1/10	16-20	
Quick	0.2	1/5	7-9	Not Required
	0.25	1/4	5-7	
	0.33	1/3	4-5	
Fast	1	1	1.2	Required
	2	2	0.6	
	4	4	0.3	
Trickle	0.02-0.1	1/50-1/10	Used for maintaining charge of a fully charged battery	

Figure 5–8 shows the charge acceptance of NiCd cells. The ideal cell, with no charge acceptance losses, would be 100 percent efficient. All the charge delivered to the cell could be retrieved on discharge. But nickel-cadmium cells typically accept charge at different levels of efficiency depending upon the state of charge of the cell, as shown by the bottom curve of Figure 5–8.

For successive types of charging behavior—Zones 1, 2, 3, and 4 in Figure 5–8 describe this performance. Each zone reflects a distinct set of chemical mechanisms responsible for loss of charge input energy.

In Zone 1 a significant portion of the charge input converts some of the active material mass into a non-usable form, i.e., into charged material which is not readily accessible during medium or high-rate discharges, particularly in the first few cycles. In Zone 2, the charging efficiency is only slightly less than 100 percent; small amounts of internal gassing and parasitic side reactions are all that prevent the charge from being totally efficient. Zone 3 is a transition region.

As the cell approaches full charge, the current input shifts from charging positive active material to generating oxygen gas. In the overcharge region, Zone 4, all of the current coming into the cell goes to generating gas. In this zone the charging efficiency is practically zero.

The boundaries between zones, 1, 2, 3, and 4 are indistinct and quite variable depending upon cell temperature, cell construction, and charge rate. The level of charge acceptance in Zones 1, 2, and 3 is also influenced by cell temperature and charge rate. For details Energy Products (1992) is suggested.

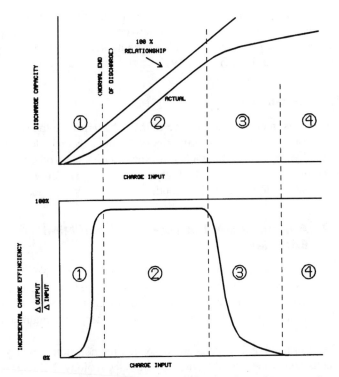

FIGURE 5-8   Charge acceptance of a sealed NiCd Cell at 0.1°C and 23°C

5.5.4 Voltage Depression Effect

When some NiCd batteries are subjected to numerous partial discharge cycles and overcharging, cell voltage decreases below 1.05 V/cell before 80 percent of the capacity is consumed. This is called the voltage depression effect and the resultant lower voltage may be below the minimum voltage required for proper system operation, giving the impression that the battery has worn out. See Figure 5–7(b). In cells exposed to overcharge, particularly at higher temperatures, this is quite common and the voltage may be about 150 mV lower than the normal cell voltage.

Voltage depression is an electrically reversible condition and disappears when the cell is completely discharged and charged. This process is sometimes called conditioning. This effect is sometimes erroneously called the "memory effect."

5.6 Nickel Metal Hydride Batteries

For those battery users who need high power in a small package and are willing to spend a higher price, there is a newer option. Battery manufacturers have started producing a newer range of batteries—Nickel-Metal Hydride

(NiMH) families which offer significant increase in cell power density. These extensions of the NiCd cell technology to new chemistry are becoming popular with product applications such as notebook computers etc. The first practical NiMH batteries entered the market in 1990.

5.6.1 Construction

In many ways, nickel-metal hydride (NiMH) batteries are the same as NiCd types, but they use nickel for the positive electrode, and a recently developed material known as a hydrogen absorbing alloy, for the negative electrode. With an operating voltage of 1.2V, they provide high-capacity, large energy density characteristics comparable to those of NiCd models.

5.6.2 A Comparison Between NiCd and NiMH Batteries

The NiCd cell is more tolerant of fast recharging and overcharging than NiMH cells are. NiCd cells hold their charge longer than do NiMH cells. NiCd cells will withstand 1000 charge/discharge cycles compared to about 500 cycles for NiMH cells. Further NiCd cells will withstand a wider temperature range than NiMH cells.

On the other hand, NiMH cells do not exhibit the notorious "memory effect" that NiCd cells sometimes do. As with any new technology, NiMH's prices are higher than those of NiCds (Small (1992) to Briggs (1994)).

Voltage profile of NiMH cells during discharging is very similar to that of the NiCd cells. NiMH cells' open-circuit voltage is 1.3 to 1.4 V. At moderate discharge rates, NiMH cells' output voltage is 1.2V. Both NiCd and NiMH cells have relatively constant output voltage during their useful service. Figure 5–9 is a typical

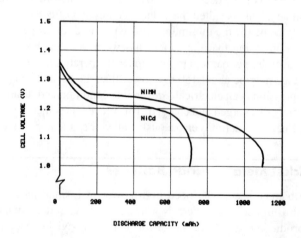

FIGURE 5-9 Comparison of discharge characteristics of NiCd and NiMH batteries

graph from a battery company comparing the output voltage of 700-mAh NiCd and 1100-mAh NiMH AA cells while under load. Note that the NiMH cell's greater capacity results in approximately 50 percent longer service life.

Figure 5-10 is another typical battery-company graph showing that NiCd and NiMH batteries and cells charge in similar fashions as well. However, the little bumps at the end of the two cells' charge curves bear closer examination. You will always see these negative excursions even though absolute cell voltages vary significantly with temperature.

Thus, the negative excursions signal a fully charged cell more or less independently of temperature, a useful quirk that sophisticated battery chargers exploit. Note that the NiCd cell's negative-going voltage excursion after reaching full charge is more pronounced than the NiMH cell's.

There are several reasons for replacing NiCd batteries with NiMH types:

- NiCd batteries contain cadmium, which is harmful to the environment.
- NiMH has nearly the same operating voltage as NiCds, making them interchangeable.
- NiMH batteries have 30 to 40 percent higher capacity for the same physical size as NiCds.
- NiMH batteries have 90 minute charging capability.

5.7 Lithium-Ion (Li-Ion) Batteries

The demand for portable systems is increasing at a dramatic rate. To remain competitive, companies are offering lighter-weight and longer run-time systems. Meeting these goals requires improvements in battery technologies beyond traditional NiCd and NiMH systems.

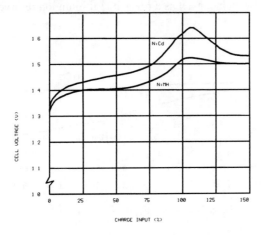

FIGURE 5-10 Battery voltage at the achievement of 100 percent charge

154 POWER ELECTRONICS DESIGN HANDBOOK

Li-Ion is a promising technology that can improve capacity for a given size and weight of a battery pack. With an energy density by weight about twice that of nickel-based chemistries (Table 5–1), Li-Ion batteries can deliver lighter weight packs of acceptable capacity. Li-Ion also has about three times the cell voltage of NiCd and NiMH batteries; therefore fewer cells are needed for a given voltage requirement.

The first noticeable difference between Li-Ion and nickel based batteries is the higher internal impedance of the lithium-based batteries. Figure 5–11 shows this by graphing the actual discharge capacity of a Li-Ion cell at different discharge currents compared to a NiCd cell.

At a 2A discharge rate (2C), less than 80 percent of the rated capacity is available for Li-Ion compared to nearly 95 percent of the rated capacity for NiCd. For systems with discharge currents greater than 1A, the capacity realized from the Li-Ion battery may be less than expected. Parallel battery stack configurations are often used in Li-Ion battery packs to help reduce the severity of this problem.

Li-Ion technology requires re-examination of the charge and discharge characteristics of portable systems. Due to the nature of the chemistry, Li-Ion batteries cannot tolerate overcharge and over-discharge.

Today, commercially available packs have an internal protection circuit that limits the cell voltage during charge to between 4.1 and 4.3V per cell, depending on the manufacturer. Voltages higher than this rating could permanently damage the cell. A discharge limit of between 2.0 and 3.0 V (depending on the manufacturer) is necessary to avoid reducing the cycle life of the battery and damaging the battery.

5.7.1 Construction

The anode, or negative electrode, in a Li-Ion cell is comprised of a material capable of acting as a reversible Li-Ion reservoir. This material is usually a form of carbon, such as coke, or graphite, or pyrolytic carbon. The cathode, or positive electrode, is also a material which can act as a reversible lithium ion reservoir.

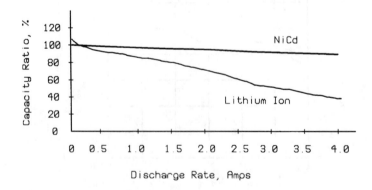

FIGURE 5–11 Li-Ion and NiCd capacity vs. discharge current

Preferred cathode materials are currently $LiCoO_2$, $LiNiO_2$, or $LiMn_2O_4$ because of their high oxidation potentials of about 4 volts versus lithium metal. Commercially available Li-Ion cells use a liquid electrolyte made up of mixtures which are predominately organic carbonates containing one or more dissolved lithium salts (Levy 1995).

5.7.2 Charge and Discharge Characteristics

Today, the predominant Li-Ion technologies use coke or graphite for an anode material. Figure 5–12 illustrates the differences in the two types of cells during discharge. The graphite anode discharge voltage is relatively flat during a majority of the discharge cycle, while the coke anode discharge voltage is more sloped (Juzcow and St. Louis (1996)).

The energy available from the graphite anode cell is higher for a given capacity due to the higher average discharge voltage. This may be useful in systems that need the maximum watt-hour capacity for a given battery size. Also, the charge and discharge cut-off voltages between the two Li-Ion systems vary among manufacturers.

Figure 5–13 shows the typical charge profile for Li-Ion batteries. The charge cycle begins with a constant current limit, transitioning to a constant-voltage limit, typically specified between 4.1 V and 4.3 V ±1%. This allows maximum charge capacity without cell damage.

Charging to a lower voltage limit does not damage the cell, but the discharge capacity will be reduced. A 100mV difference could change the discharge capacity by more than 7 percent. Basically, the difficult aspect of this type of charger is the wide dynamic range required from the switching current regulator given the tight

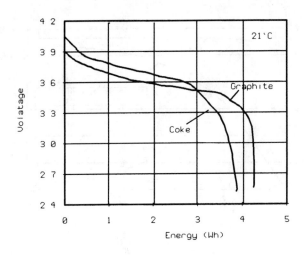

FIGURE 5–12 Li-Ion discharge profile for different electrodes (Courtesy of Moil Energy Ltd., USA)

voltage tolerance. Some chargers provide single-cell monitoring, while others rely on the internal protection circuit to do this.

5.8 Reusable Alkaline Batteries

Alkaline technology has been used in primary batteries for several years. With the development of the reusable alkaline manganese technology, secondary alkaline cells are quickly making their way into many consumer and industrial applications. Under many applications, reusable alkaline cells can be recharged from 75 to over 500 times, unlike single-use alkaline batteries, and initially have three times the capacity of a fully charged NiCd battery. These cells do not, however, compete with NiCds in high-power applications.

Battery Technologies Inc. (BTI) developed the rechargeable (or reusable) alkaline manganese technology and has licensed its technology to several companies throughout the world. For example, Rayovac Corporation, one of the licensees, launched its line of reusable alkaline products under the name RENEWAL (Nossaman and Pavershi (1996) and Sengupta (1995)).

The chemistry behind the reusable approach depends on limiting the zinc anode to prevent over-discharge of the MnO_2 cathode. Additives are also incorporated to control hydrogen generation and other adverse effects on charge. Rated cycle life is 25 cycles to 50 percent of initial capacity. Longer cycle life is possible depending on drain rate and depth of discharge. To take advantage of the reusable alkaline cell and increase its life, a special "smart charger" is required.

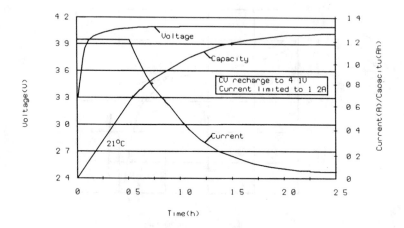

FIGURE 5-13 Li-Ion charge profile at constant potential charging at 4.1 V and current limited to 1.2A (Courtesy of Moli Energy Ltd., USA)

5.8.1 Cumulative Capacity

Using reusable alkaline cells can drive down the total battery cost to the consumer. This cost savings can be determined by looking at the cumulative capacity of a reusable cell versus the one-time use of a primary alkaline cell. Figure 5–14 illustrates the capacity of AA cells being discharged down to 0.9V at 100mA. It shows that although the initial use of the reusable alkaline is almost that of primary alkaline, the reusable one can be recharged for continued use. Table 5–2 shows the increase in cumulative capacity by limiting the DOD and achieving more cycles.

TABLE 5–2 Capacity of "AA" Cells at Various Depths of Discharge (DOD) (Values in mAh) (Courtesy: Battery Technologies Inc., USA)

Condition	Capacity/Cumulative Capacity at 125mA discharge rate up to 0.9V		
	100% DOD	30% DOD	10% DOD
Cycle 1	1,500	450	150
Cycle 50	400		
Cumulative 50	33,000	22,000	7,000
Cumulative 100		44,000	15,000
Cumulative 500			73,000

Overcharging also affects the cycle life of reusable alkaline. Reusable alkaline is not tolerant of overcharge and high continuous charge currents, and may be damaged if high current is forced into them after they have reached a partially recharged state. Proper charging schemes should be used to prevent an overcharged condition.

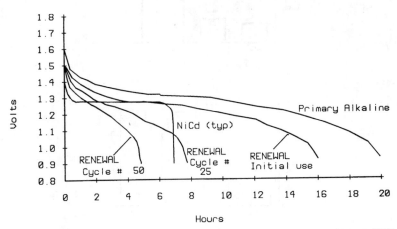

FIGURE 5–14 100 mA discharge curve comparison for NiCd, primary alkaline and reusable alkaline (Source: Nossaman and Parvereshi 1996)

5.9 Zn-Air Batteries

Primary Zn-air batteries have been in existence for over 50 years with applications such as hearing aids and harbor buoys. The light weight and high energy content in Zn-air technology has promoted research on Zn-air rechargeable chemistry by companies such as AER Energy Resources, USA. Initial focus was on electric vehicles, which later shifted towards portable appliances.

Rechargeable Zn-air technology is an air breathing technology where the oxygen in ambient air is used to convert zinc into zinc oxide in a reversible process. Cells use air breathing carbon cathode to introduce oxygen from air into potassium-hydroxide electrolyte. Cathode is multi-layered with a hydrophilic layer and anode is comprised of metallic zinc.

The characteristic voltage of the rechargeable Zinc-air system is a nominal 1 volt. During discharge, the cells will operate at a voltage between 1.2 and 0.75 V. The current and power capability of the system is proportional to the surface area of the air breathing cathode. For more current and power, a larger surface area cell is required. For less current and power, a smaller cell may be used. Compared to other rechargeable chemistries, Zn-air needs an air manager for an intake and exhaust of air to allow the chemical process.

Figure 5–15 compares the gravimetric and volumetric energy densities of several rechargeable battery technologies. This clearly indicates that Zn-air batteries require less weight and volume. Discharge and charge characteristics of Zn-air batteries are shown in Figure 5–16.

(a)
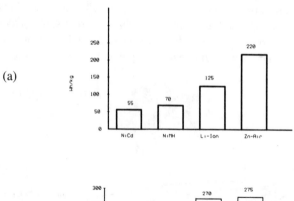

(b)
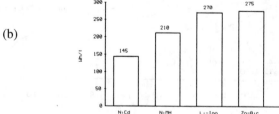

FIGURE 5–15 Energy density comparison of batteries (Courtesy: AER Energy Technologies) (a) Gravimetric (b) Volumetric

The cells exhibit a flat voltage profile over the discharge cycle. Typical charge voltage is 2 volts per cell using a constant voltage/current taper approach. Life cycle varies between 50-400 depending on the depth of discharge. Cost per Wh is apparently the lowest compared to Ni and Li based chemistries.

5.10 Battery Management

Two decades ago, battery management was having a reliable, fast, and safe charging methodology to be selected for a battery bank, together with the monitoring facilities for detecting the discharged condition of the battery pack. With modern battery technologies emerging, the demands from the cost sensitive portable product market, as well as the medium power range products such as UPS and telecom power units, etc., attributes of a modern battery management system include:

- Battery charging methods and charge control
- End of discharge determination
- Gas gauging

(a)
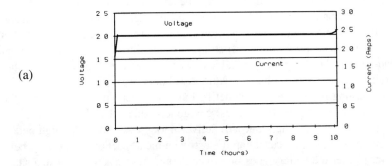

(b)
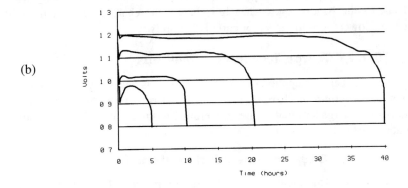

FIGURE 5-16 Charge/discharge characteristics of Zn-air chemistry (Courtesy: AER Energy Technologies) (a) Charge (b) Discharge

- Monitoring battery health issues
- Communication with the host system/or power management sub-systems.

The following sections provide some concepts and techniques related to managing nickel based chemistries, sealed-lead acid, and Li-Ion battery systems. A discussion on rechargeable alkaline batteries beyond the depth of this chapter. Kuribayashi (1993) and Levy (1995) provide some details on managing rechargeable alkaline cells.

5.10.1 Charging Systems

While four major battery families can accept either a standard (16 to 24 hour) or a fast (2 to 4 hour) charge, discussion here is limited to fast charging methods. Slower charging schemes tend to be found in simpler, price sensitive applications, which do not need (or cannot afford) much beyond a charger and a low battery indicator.

The objective of fast charging a battery is to cram as much energy as it takes to bring the battery back to fully charged state in the shortest possible time without damaging the battery or permanently affecting its long term performance. Since current is proportional to energy divided by time, the charging current should be as high as the battery systems will reasonably allow. For the constant-current cells (NiCd and NiMH), a 1C charge rate will typically return more than 90 percent of the battery's useable discharge capacity within the first hour of charging. The constant-potential cells (Lead-Acid and Lithium-Ion) are a bit slower to reach the 90 percent mark, but can generally be completely recharged within five hours.

Fast-charging has compelling benefits, but places certain demands upon the battery system. A properly performed fast-charge, coordinated to the specifications of a battery rated for such charging, will deliver a long cycle life. The high charging rates involved, however, cause rapid electrochemical reactions within the cells of the battery. After the battery goes into overcharge, these reactions cause a sharp increase in internal cell pressure and temperature.

Uncontrolled high-rate overcharge quickly causes irreversible battery damage. Thus, as the battery approaches full charge, the charging current must be reduced to a lower "top-off" level, or curtailed entirely.

5.10.1.1 Charge Termination Methods

If a rapid charge is applied to a battery pack, it is necessary to select a reliable method to terminate charging at the fully charged position. Two practical approaches for charge termination are temperature termination method and voltage termination method.

5.10.1.1.1 Temperature Termination Method

Temperature is the main cause of failure in a rechargeable cell, so it makes sense to monitor the cell temperature to determine when to shut off charge to a battery. Three methods of charge termination, based on temperature, are common: maximum temperature cutoff (MTC), temperature difference (ΔT), and temperature slope (dT/dt). The maximum temperature cutoff system is the easiest and cheapest to implement, but is the least reliable. Using a bimetallic thermal switch or a positive temperature coefficient thermistor, a simple, low cost circuit can shut down a charging current at an appropriate temperature.

The temperature difference (ΔT) method measures the difference in ambient and cell temperatures to compensate for a cool environment. The ΔT method requires monitoring two temperature sensors, one for the battery temperature and one for the ambient. This method may be unsuitable if the difference between cell and ambient temperature is very large.

The ΔT method can become unreliable with a quickly changing ambient temperature unless an equal thermal mass is attached to the ambient sensor. This means that ΔT method is suitable for a primary charge termination at lower charge rates up to C/5 if the ambient temperature is not going to change often. The ΔT method also provides an excellent backup charge termination scheme.

The temperature slope method (dT/dt), a more sophisticated temperature termination scheme, measures the change in temperature over time. This method uses the slope of the battery temperature curve, and therefore is less dependent on changes in ambient temperatures or in large differences between ambient and battery temperatures. Accurately adjusted to a particular pack, and with careful attention paid to the type and placement of the temperature sensor, the dT/dt method works very well. This dT/dt method is suitable for charge rates up to 1C, and provides an excellent backup method.

5.10.1.1.2 Voltage Termination Methods

Four commonly available voltage termination methods are maximum voltage (V_{max}), negative delta voltage ($-\Delta V$), zero slope, and inflection point (dV/dt). The maximum voltage method senses the increase in battery voltage as the battery approaches full charge. However, this is accurate only on a highly individualized basis. It is necessary to know the exact value at the voltage peak, otherwise the batteries may be over or undercharged.

Temperature compensation is also required because of the negative temperature coefficient of battery voltage. The maximum voltage will increase if the batteries are cold, causing an undercharge because the charging voltage will reach the maximum voltage trip point early. If the batteries are hot, the maximum voltage may never be reached and the batteries will be cooked. Therefore, the V_{max} method is generally not recommended for fast charge rates.

The negative delta voltage is the most popular of the fast charge termination schemes. It relies on the characteristic drop in cell voltage that occurs when a battery enters overcharge, as shown in Figure 5–17. With most NiCd cells, the voltage drop is a very consistent indicator, and the $-\Delta V$ method is fine for charge rates up to 1C. An inherent problem with this method is that the batteries must be driven into the overcharge region to cause the voltage decrease. Pressures and temperatures rise very rapidly at fast charge rates beyond 1C. In cyclic applications the battery must be able to endure that continual abuse.

Another concern is that cells like NiMH types do not always have the characteristic decrease in voltage, compared to NiCds as shown in Figure 5–10. This creates a problem of forward compatibility when moving from NiCd to NiMH cells. Most manufacturers of NiMH cells do not advocate the $-\Delta V$ method of charge termination.

The zero slope method monitors the point where the slope of the battery voltage reaches zero. This method is reliable for rapid charge rates up to 4C, and is less susceptible to noise on the voltage sense lines. However, a few types of batteries such as the NiCd button cells etc. may have a voltage slope that never quite reaches zero. Therefore, the zero slope method is better suited as a backup method.

In the inflection point (dv/dt) method the system monitors the change in voltage over time, and is the most sensitive indicator for preventing overcharge. The inflection point method relies on the changes in the voltage slope, shown in Figure 5–17, which occurs during charge and is an excellent primary termination method for up to 4C charge rates.

The change in the voltage slope is an extremely reliable and repeatable indication of charge. It does not rely on the decrease in voltage that may not always occur. Instead, this method looks for the flattening of the voltage profile as the battery reaches full charge. By monitoring the relative change in the steepness of the voltage slope, this method avoids having to use absolute numbers.

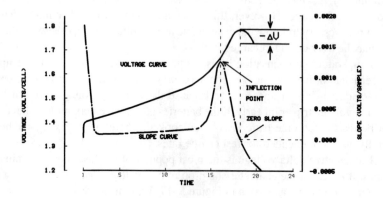

FIGURE 5–17 Termination methods based on changes in voltage and voltage slope

5.10.1.2 NiCd and NiMH Fast Charge Methods

Nickel based batteries such as NiCd and NiMH types have become the most popular choice for the portable products. Although it is not correct to consider the NiCd and NiMH electrochemistries or charging regimens as being interchangeable, they are similar enough that they can be discussed together.

There is no one best way to fast-charge a NiCd or NiMH battery. Variables introduced by the allowable cost and size of the end application, the choice of charge termination method(s), and the specific battery vendor's recommendations will all influence the final choice of charging technique.

(a)
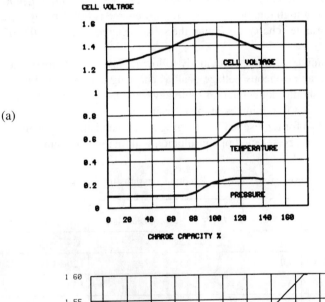

(b)
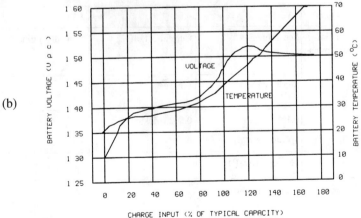

FIGURE 5-18 Charging indications for nickel based batteries at 1C charge rate (a) Cell voltage, temperature and pressure for a typical NiCd cell at 1C rate charging (b) NiMH voltage and temperature characteristics

Figure 5–18(a) shows the voltage, pressure, and temperature characteristics of a NiCd cell being charged at the 1C rate. Figure 5–18(b) shows similar data for a NiMH cell.

These curves illustrate the need for a reliable termination of the high-current portion of the charge cycle, and assist in understanding the various fast-charge termination methods outlined in Table 5–3. For both electrochemistries, the ideal fast-charge termination point is at 100 percent to 110 percent of returned charge. The charging current is then reduced to the top-off value for one to two hours, to bring the cell into a state of slight over charge.

This compensates for the inefficiencies of the charging process (e.g., heat generation). If the specific application will have the battery on standby for more than several weeks, or at high temperatures, the top-off charge is followed by a continuous, low-level "trickle" charge to counter the self-discharge characteristics of NiCd and NiMH cells.

Under certain conditions, particularly following intervals of storage, a NiMH battery may give an erroneous voltage peak as charging commences. For this reason, the charger should deliberately disable any voltage-based charge termination technique for the first five minutes of the charging interval.

See Table 5–3 for representative charging recommendations. The most appropriate method should be selected in consultation with the manufacturer.

TABLE 5–3 Representative Charging Recommendations for Different Batteries

	NiCd or NiMH	Sealed Lead Acid	Li-Ion
Charging Current	1C	1.5C	1C
Voltage per Cell (Volts)	1.80	2.5	4.20 ± 0.05
Charging time (hours)	~3	~3	2.5 to 5.0
Method for optimum fast-charge termination point	See Table 5–4	Current cut off	Typically a timer[a]
Back-up charge termination method	See Table 5–4	Timer	
"Top off" rate	0.1C	0.002 (trickle)	
Temperature range (°C)	10°-40° (NiCd) 150°-30° (NiMH)	0-30°	0-40°

[a] Depending on manufacturer's recommendation

Table 5–4 summarizes the fast charge termination methods for NiCd and NiMH cells.

TABLE 5-4 Fast Charge Termination Methods for NiCd and NiMH Batteries

Charging Technique	Description
Negative ΔV ($-\Delta V$)	Looks for the downward slope in cell voltage which a cell exhibits ($\approx$ 30 mV to 50mV for NiCd, 5mV to 15mV for NiMH) upon entering overcharge. Very common in NiCd applications due to its simplicity and reliability.
Zero ΔV	Waits for the time when the voltage of the cell under charge stops rising, and is "at the top of the curve" prior to the downslope seen in overcharge. Sometimes preferred over $-\Delta V$ for NiMH, due to NiMH's relatively small downward voltage slope.
Voltage Slope (dV/dt)	Looks for an increasing slope in cell voltage (positive dV/dt) which occurs somewhat before the cell reaches 100% returned charge (prior to the Zero ΔV point).
Inflection Point Cutoff (d_2V/dt_2, IPCO)	As a cell approaches full charge, the rate of its voltage rise begins to level off. This method looks for a zero or, more commonly, slightly negative value of the second derivative of cell voltage with respect to time.
Absolute Temperature Cutoff (TCO)	Uses the cell's case temperature (which will undergo a rapid rise as the cell enters high-rate overcharge) to determine when to terminate high-rate charging. A good backup method, but too susceptible to variations in ambient temperature conditions to make a reliable primary cutoff technique.
Incremental Temperature Cutoff (Δ TCO)	Uses a specified increase in the cell's case temperature, relative to the ambient temperature, to determine when to terminate high-rate charging. A popular, relatively inexpensive and reliable cutoff method.
Delta Temperature/Delta Time ($\Delta T/\Delta t$):	Uses the rate of increase of a cell's case temperature to determine the point at which to terminate the high rate charge. This technique is inexpensive and reliable once the cell and its housing have been properly characterized.

5.10.1.3 Sealed Lead-Acid Batteries

Unlike nickel based batteries, sealed lead-acid (SLA) batteries are charged using a "Constant-Potential" (CP) regimen. CP charging employs a voltage source with a deliberately imposed current limit (a current-limited voltage regulator). A significantly discharged battery undergoing CP charging will initially attempt to draw a high current from the charger. The current limiting function of the CP regimen serves to keep the peak charging current within the battery's ratings.

Following the current limited phase of the charging profile, a SLA battery exposed to a constant voltage will exhibit a tapering current profile as shown in

Figure 5–19. When returned charge reaches 110 percent to 115 percent of rated capacity, allowing a dischargeable capacity of 100 percent of nominal, the charge cycle is complete.

The specifics of fast-charging SLA batteries are more vendor-dependent than those of NiCd or NiMH units. The information in Table 5–3 uses data from GS battery (USA) Inc. The primary termination method, "current cutoff," looks at the absolute value of the average charging current flowing into the battery.

When that current drops below 0.01C, the battery is fully charged. If it will be in standby for a month or more, a trickle current of 0.002C should be maintained. The backup termination method, according to the vendor's recommendations, should be a 180 minute timeout on the charging cycle (Schwartz 1995).

To satisfy more stringent charge control recommendations, where the battery temperature, voltage, and current need be sampled, many dedicated charge controller ICs are available in the market. Bq2031 lead acid fast charge IC from Benchmarq Microelectronics Inc. and UC2906 (sealed-lead acid charger) from Unitrode Integrated Circuits are examples.

The UC2906 battery charger controller contains all of the necessary circuitry to optimally control the charge and hold cycle for sealed-lead acid batteries. These integrated circuits monitor and control both the output voltage and current of the charger through three separate charge states; a high current bulk-charge state, a controlled over-charge, and a precision float-charge, or standby state. Figure 5–20 shows a block diagram and one implementation of UC2906 in a dual step current charger. Sacarisen and Parvereshi (1995) and Unitrode Inc. (1995–1996) provide details about sealed-lead acid charge control techniques.

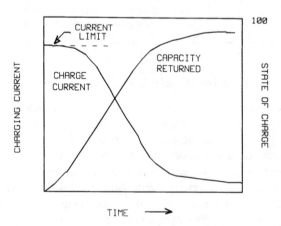

FIGURE 5–19 Typical current and capacity returned v̇s. charge time for CP charging

5.10.1.4 Li-Ion Chargers

Lithium-Ion batteries require a constant potential charging regimen, very similar to that used for lead-acid batteries. Typical recommendations for Li-Ion fast-charge are indicated in Table 5–3. As with lead-acid batteries, a Li-Ion cell under charge will reduce its current draw as it approaches full charge.

If the cell vendor's recommendation for charging voltage (generally 4.20V ±50mV at 23°C) is followed, the cells will be able to completely recharge from any "normal" level of discharge within five hours. At the end of that time, the charging voltage should be removed. Trickle current is not recommended.

(a)
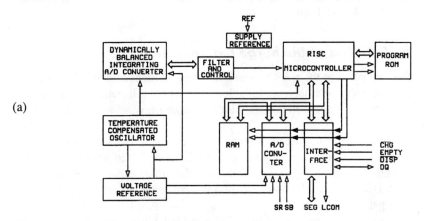

(b)
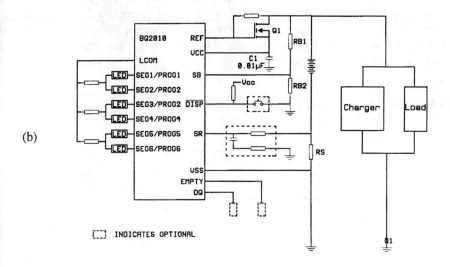

FIGURE 5–20 UC2906 and its implementation (Reproduced by permission of Unitrode Inc., USA) (a) Block diagram (b) Implementation in a dual step current charger

If the voltage on a Li-Ion cell falls below 1.0V, recharging of that cell should not be attempted. If the voltage is between 1.0V and the manufacturer's nominal minimum voltage (typically 2.5V to 2.7V), it may be possible to salvage the cell by charging it with a 0.1C current limit until the voltage across the cell reaches the nominal minimum, followed by a fast-charge.

Due to special characteristics of Li-Ion batteries, most Li-Ion manufacturers incorporate custom circuits into their battery packs to monitor the voltage across each cell within the battery and to provide protection against overcharge, battery reversal, and other major faults. These circuits are not to be confused with charging circuits. For example the MC 33347 protection circuit is such a monolithic IC from Motorola (Alberkrack, 1996).

5.10.2 End of Discharge Determination

Determination of the point at which a battery has delivered all of its usefully dischargeable energy is important to the longevity of the cells which form that battery. Discharging a single cell too far will often cause irreversible physical damage inside the cell.

If multiple cells are placed in series, unavoidable imbalances in their capacities can cause the phenomenon known as "cell reversal," in which the higher-capacity cells force a backward current through the lowest-capacity cell. Knowing the End Of Discharge (EOD) point provides a "zero capacity" reference for coulometric gas gauging.

The actual determination of the EOD point is typically done by monitoring cell voltage. For the most accurate determination of EOD when the load is varying, correction factors for load current and the battery's state of charge should be applied, especially to Lead-Acid and Li-Ion batteries. The essentially flat discharge profiles of NiCd and NiMH make these corrections a matter of user discretion for most load profiles. Table 5–5 shows voltages commonly used to indicate the End Of Discharge point for the four battery types.

TABLE 5–5 Typical End of Discharge Voltages

Cell Type	EODV (Volts)	Comments
Lead Acid	1.35 to 1.9 V (1.8 V Typical)	Dependent upon loading, state of charge, cell construction and manufacturer
NiCd	0.9	Essentially constant
NiMH	0.9	Essentially constant within recommended range of discharge rates
Li-Ion	2.5 to 2.7	Dependent upon manufacturer, loading and state of charge

5.10.3 Gas Gauging

The "gas gauging" concept discussed here does not refer to the gases which may be evolved by the battery reactions, but rather to the concept of the battery as a fuel tank powering the product. Gas gauging, therefore, involves real time determination of a battery's state of charge, relative to the battery's nominal capacity when fully charged.

It is possible to make an inexpensive and moderately useful state of charge measurement from a simple voltage reading, if the battery being used has a sloping voltage profile. Hence, Li-Ion batteries and, to a lesser extent, SLA batteries, should be amenable to such an approach. In practice, the results are less than optimal: cell voltages are dependent upon loading, internal impedance, cell temperature, and other variables.

This reduces the attractiveness of the battery voltage method of gas gauging; the fact that it is not suitable for NiCd or NiMH batteries, due to their essentially flat voltage profile, makes it commercially untenable. A clever and effective alternative is the "coulometric" method.

Coulometric gas gauging, as its name implies, meters the actual charge (current times time) going into and out of the battery. By integrating the difference of current in and current out, it is possible to determine the charge status of the battery at any given time. There are of course real-world details which must be observed in the actual implementation of such a gas gauge; some of the most important of these are:

(a) It is necessary to have an accurate starting point for the integrator, corresponding to a known state of charge in the battery. This is often resolved by zeroing the integrator when the battery reaches its EODV.
(b) Compensation for the temperature. The actual capacity of lead-acid batteries increases with temperature, that of nickel based batteries decreases as battery temperature rises.
(c) Appropriate conversion factors should be applied for the particular charge regimen and discharge profile used. Under conditions of highly variable battery loading, dynamic compensation may be advisable.

5.10.4 Battery Health

With battery health defined as a battery's actual capacity relative to its rated capacity, the health of the battery can be determined and maintained in three steps:

i. Discharge the battery to the EOD point, preferably into a known load
ii. Execute a complete charge cycle, while gas gauging the battery
iii. Compare the battery's measured capacity to its rated capacity.

This sequence will simultaneously "condition" the battery (which means to overcome the so-called "memory effect" of capacity of NiCd batteries, for example), and will indicate the capacity of the battery after conditioning. The information so obtained can be used to ascertain whether the battery is in good shape, or is approaching the end of its useful life.

5.10.5 Semiconductor Components for Battery Management

In practical systems charging of rechargeable batteries could be done by a variety of components such as simple voltage regulator ICs to microprocessor based systems (Kerridge 1993). Specially designed battery-management ICs provide fine control and can also give a useful indication of a battery's charge condition both on and off charge.

Battery voltage, charge/discharge current, and cell case temperature are the clues available to battery-management ICs. However, there is one difficulty: each clue is the result of a chain of events far removed from the chemical reaction which is necessary to monitor and control. Further complexities arise because the three properties interact. Because instantaneous measurements of any of the three attributes are virtually useless, all battery management ICs include some sort of time to add meaning to the data. Table 5–6 shows some representative battery management ICs from different manufacturers.

TABLE 5–6 Battery Management Integrated Circuits

Manufacturer	Part Number	Description	Cell Type				
			NiCd	NiMH	Li-Ion	Lead Acid	Recharge Alkaline
Benchmarq Micro Electronics	bq 2003	Fast charge IC	Yes	Yes	-	-	-
	bq 2010	Gas Gauge IC	Yes	Yes	-	-	-
	bq 2040	Gas gauge IC with SM bus Interface	Yes	Yes	Yes	-	-
	bq 2031	Fast charger IC	-	-	-	Yes	-
	bq 2053	Li-Ion Pack supervisor	-	-	Yes	-	-
	bq 2902	Charge/Discharge controller	-	-	-	-	Yes
Unitrode Integrated Circuits	UC 3905	Charge Controller	Yes	Yes	-	-	-
	UC 3906	Charger	-	-	-	Yes	-
	UC 3909	Switchmode charger	-	-	-	Yes	-
Dallas Semi-conductor	DS 1633	High speed charger	Yes	Yes	Yes	Yes	Yes
	DS 1837	Quick Battery recharge	Yes	Yes	Yes	Yes	Yes

There are many means of providing a complete universal battery management system. Years ago, most companies involved with design of chargers, etc. for battery banks used to have design resources dedicated internally to provide these types of solutions. Today, with the rapid changes in rechargeable battery chemistries and the associated charge/discharge management requirements, many companies prefer to use standardized semiconductor components dedicated to this function.

Two possible approaches involve using a customized microcontroller or an ASIC dedicated to battery management. Figure 5–21 shows the block diagram of a microcontroller-based universal battery monitor.

Using ASIC solutions, such as bq 2010 from Benchmarq Microelectronics, standby current necessary for battery monitoring circuits can be minimized, while providing many gas gauge functions. A block diagram of bq 2010 is shown in Figure 5–22(a).

Specialized component manufacturers have recently introduced many gas gauge ICs for multi chemistry (NiCd/NiMH/Li-Ion) environments (Freeman (1995) and Benchmarq (1997)).

VLSI components such as bq 2040 etc. with the SMBus interface (discussed later) are intended for battery-pack or in-system installation to maintain an accurate record of available battery charge. In addition to supporting SMBus protocol, it supports Smart Battery Data (SBData) specification (Rev 1.0) (Dunstan 1995). See Figure 5–23.

Another example is Linear Technology Corporations LTC^R 1325 battery management IC which provides a complete system which can accommodate all plausible electrochemistries, and their concomitant charging needs and charge termination algorithms, with few or no hardware changes for different battery types.

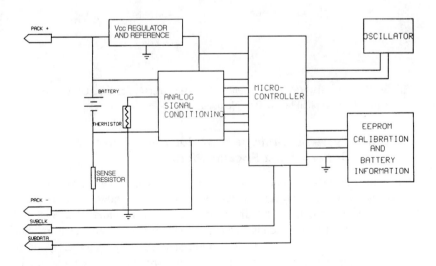

FIGURE 5–21 Microcontroller based universal battery monitor

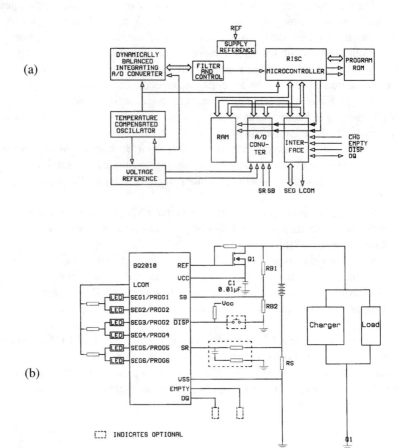

FIGURE 5-22 bq 2010 (Courtesy: Benchmarq Microelectronics Inc., USA) (a) Block diagram of bq 2010 (b) Typical bq 2010 implementation

5.10.6 Systems Management Bus and Smart Battery Data Specifications

Given the methods available, new industry standards are being proposed that standardize the battery and power management subsystems within portable products. One standard proposed by Intel and Duracell includes the Systems Management Bus (SMBus), which is a standardized means of communicating with the various subsystems within a portable system. Smart Battery Data (SBD) specification outlines the type of data that a battery must provide the system so it can properly implement power and battery management. Dunstan (1995) provides details on SMBus and related Application Programming Interface (API).

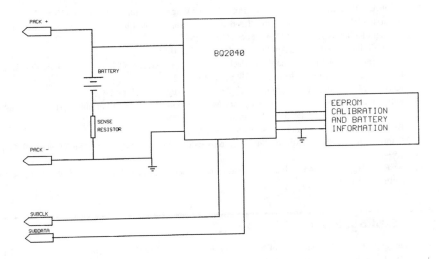

FIGURE 5-23 The implementation of a universal battery monitor using bq 2040 (Courtesy: Benchmarq Microelectronics, Inc.)

References

1. Energy Products Inc. *Rechargeable Batteries Applications Handbook*. Butterworth-Heinemann, 1992.
2. Hirai, T. "Sealed Lead-Acid Batteries find Electronic Applications." *PCIM*, January 1990, pp 47–51.
3. Moore, M. R. "Valve Regulated Lead Acid Vs Flooded Cell." Power Quality Proceedings, October 1993, pp 825–827.
4. Moneypenny, G.A. and F. Wehmeyer. "Thinline Battery Technology for Portable Electronics." HFPC Conference Proceedings, April 1994, pp 263–269.
5. Small, C. H. "Nickel-Hydride Cells Avert Environmental Headaches." *EDN*, 10 December 1992, pp 156–161.
6. Eager, J. S. "The Nickel-Metal Hydride Battery." Power Conversion International Proceedings, September 1991, pp 398–404.
7. Furukawa, N. "Developers Spur Efforts to Improve Ni-Cd, Ni-MH Batteries." *JEE*, October 1993, pp 46–50.
8. Briggs, A: "Ni-MH Technology overview"; Portable by Design Conference Proceedings, 1994, pp BT-42 to BT-45.
9. Kuribayashi, I. "Needs for Small Batteries Spur Progress in Lithium-Ion Models." *JEE*, October 1993, pp 51–54.
10. Levy, S. C. "Recent Advances in Lithium Ion Technology." Portable by Design Conference Proceedings, 1995, pp 316–323.
11. Juzkow, M. W. and C. St. Louis. "Designing Lithium-Ion Batteries Into Today's Portable Products." Portable by Design Conference, 1996, pp 13–22.
12. Nossaman, P. and J. Parvereshi. "In Systems Charging of Reusable Alkaline Batteries." Proceedings of Portable by Design Conference (USA), March 1996.

13. Sengupta, U. "Reusable Alkaline™ Battery Technology: Applications and System Design Issues for Portable Electronic Equipment." Portable by Design Conference, 1995, pp 562–570.
14. Schwartz, P. "Battery Management." Portable by Design Conference, 1995, pp 525–547.
15. Sacarisen, P. S. and J. Parvereshi. "Lead acid fast charge controller with improved battery management techniques." South Conference, March 1995.
16. Unitrode Inc. "Product and Applications Handbook." 1995–1996.
17. Alberkrack, J. "A Programmable In-Pack Rechargeable Lithium Cell Protection Circuit." HFPC Conference Proceedings, September 1996, pp 230–237.
18. Kerridge, B. "Battery Management Ics." *EDN Asia,* August 1993, pp 38–50.
19. Freeman, D. "Aspects of Universal Battery Management in Advanced Portable Systems." Portable by Design Conference, 1995, pp 515–524.
20. Benchmarq Microelectronics. *1997 Databook.* 1997.
21. Dunstan, R. A. "Standardized Battery Intelligence." Portable by Design Conference, 1995, pp 239–244.
22. AER Energy Resources, Inc. "Rechargeable Zinc-Air Technology Overview." Information Bulletin, May 1997.

Bibliography

1. Quinnell, R. A. "The Business of Finding the Best Battery." *EDN,* 5 December 1991, pp 162–166.
2. Harold, P. "Batteries Function in High Temperature Environments." *EDN,* 9 July 1987, pp 232.
3. Swager, A. W. "Fast Charge Batteries." *EDN,* 7 December 1989, pp 180–188.
4. Kovacevic, B. "Microprocessor-Based Nickel Metal Hydride Battery Charger Prevents Overcharging." *PCIM,* March 1993, pp 17–21.
5. Falcon, C. B. "Fast Charge Termination Methods for NiCd and NiMH Batteries." *PCIM,* March 1994, pp 10–18.
6. Heacock, D. and D. Freeman. "Single Chip IC Gauges NiMH or NiCd Battery Charge." *PCIM,* March 1994, pp 19–24.
7. McClure, M. "Energy Gauges Add Intelligence to Rechargeable Batteries." *EDN,* 26 May 1994, pp 125.
8. Cummings, G., D. Brotto, and J. Goodhart. "Charge Batteries Safely in 15 Minutes by Detecting Voltage Inflection Points." *EDN,* 1 September 1994, pp 89–94.
9. Swager, A. W. "Smart Battery Technology—Power Management's Missing Link." *EDN,* March 2, 1995, pp 47–64.
10. Duley, R. and H. David. "Battery Management Techniques for Portable Systems." Portable by Design Conference, 1994, pp BT-1 to BT-11.
11. Freeman, D. and D. Heacock. "Lithium-Ion Battery Capacity Monitoring Within Portable Systems." HFPC Conference Proceedings, 1995, pp 1–8.
12. Bowen, N. L. "System Considerations for Lithium-Ion Batteries." Portable by Design Conference Proceedings, 1996, pp 179–191.

CHAPTER **6**

Protection Systems for Low Voltage, Low Power Systems

6.1 Introduction

Modern electronic systems are uniquely vulnerable to power line disturbances because they bring together the high energy power line and sensitive low power integrated circuits. The term power conditioning is used to describe a broad class of products designed to improve or assure the quality of the AC voltage you connect to your computer, word processor, automated laboratory instrument, CAD/CAM system, telecommunications system, point-of sale terminal or other sensitive electronic systems.

Utilities realize that different types of customers require different levels of reliability, and make every effort to supply disturbance-free power. However, normal occurrences, most of which are beyond control (such as adverse weather, vehicles running into poles and equipment malfunction) make it impossible to provide disturbance-free power 100 percent of the time. In addition to these external disturbances, sources within buildings, such as switching of heavy equipment loads, poor wiring, overloaded circuits, and inadequate grounding, can cause electrical disturbances. Many of these power disturbances can be harmful to electronic equipment. Power disturbances can cause altered or lost data and sometimes equipment damage which may, in turn, result in lost production, scheduling conflicts, lost orders, and accounting problems.

There are methods and devices available to prevent these disasters from happening. Protective devices range from those providing minimal protection to those that construct a new power source for critical loads, converting the standard "utility grade power," which is adequate for most equipment, into "electronic grade power" required by some critical loads.

6.2 Types of Disturbances

6.2.1 Voltage Transients

Voltage transients are sharp, very brief increases in electrical energy. Spikes are commonly caused by the on and off switching of heavy loads such as air conditioners, electric power tools, business machines, and elevators. Lightning can cause even larger spikes. Although they usually last less than 200 microseconds, spikes can be dangerous to unprotected equipment, with amplitudes ranging from 200V to 2,500V, positive or negative, and sometimes as high as 6,000V. This high magnitude of sudden voltage variation can wipe out stored data, alter data in progress, and cause electronic hardware damage.

6.2.2 Voltage Surges

Voltage surges are voltage increases which typically last from 15 milliseconds to one-half of a second. Surges are commonly caused by the switching of heavy loads and power network switching. Surges don't reach the magnitude of sharp spikes, but generally exceed the normal line voltage by about 20 percent. This deviation can cause common computer data loss, equipment damage, and erroneous readings in monitoring systems. A surge that lasts for more than two seconds is typically referred to as an *overvoltage*.

6.2.3 Voltage Sags

Voltage sags are undervoltage conditions, which also last from 15 milliseconds to one-half of a second. Sags often fall to 20 percent below nominal voltage and are caused when large loads are connected to the power line. Sags can cause computer data loss, alteration of data in progress, and equipment shutdown. A sag that lasts for more than two seconds is typically referred to as an *undervoltage*.

6.2.4 Electrical Noise

Electrical noise is a high-frequency interference in the frequency spectrum of 7000Hz to 50MHz. Noise can be transmitted and picked up by a power cord acting as an antenna or it can be carried through the power line. These disturbances can be generated by radio frequency interference (RFI) such as radio, TV, and microwave transmission, radar, arc welding, and distant lightning. Noise can also be caused by electromagnetic interference (EMI) produced by heaters, electric typewriters, air conditioners, coffee makers, and other thermostat-controlled or motor-operated devices.

Although generally not destructive, electrical noise can sometimes pass through a power supply as if it were a signal and wipe out stored data or cause erroneous data output. Problems result when microelectronic circuitry is invaded by transient, high-frequency voltages collectively called "line noise" which can be grouped into one of two categories; normal-mode or common-mode.

6.2.4.1 Normal-Mode Noise

Normal-mode noise (sometimes referred to as transverse-mode or differential-mode noise) is a voltage differential, or potential, that appears briefly between the live wire and its accompanying neutral. As the name implies, these two lines represent the normal path of power through electric circuits, which gives any normal-mode transient a direct route into sensitive components and therefore the opportunity to destroy or degrade those components.

At today's levels of semiconductor density and sensitivity, normal-mode transient voltages can start causing degradation at around 10 volts, and can cause destruction at 40 volts. These levels may well drop further as semiconductor density increases.

6.2.4.2 Common-Mode Noise

Common-mode noise is a similar brief voltage differential that appears between the ground and either of the two supply lines. Common-mode transients are most often the cause of disruption, because digital logic is either directly or capacitively tied to the safety ground as a zero-voltage data reference point for semiconductors. As a result, transient common-mode voltage differences as small as 0.5V can cause that reference point to shift, momentarily "confusing" the semiconductors.

6.2.5 Blackouts

Blackouts result in a total loss of power. Power outages can last from several milliseconds to several hours. Blackouts are caused by many circumstances including vehicle, animal, or personal contact with power lines and equipment, tripped circuit breakers, equipment failure, and adverse weather.

6.3 Different Kinds of Power Protection Equipment

The power protection equipment can be basically divided into two primary types. They are power enhancement equipment and power synthesis equipment.

6.3.1 Power Enhancement Equipment

Power enhancement equipment modifies and improves the incoming waveform by clipping, filtering, isolating and increasing or decreasing voltage. The common types of power enhancement equipments are:

(a) Transient voltage surge suppressors (TVSS)
(b) Voltage regulators
(c) Isolation transformers
(d) Line conditioners

6.3.1.1 Transient Voltage Surge Suppressors

One or two decades ago, transients were thought to be non-existent, not to be concerned with, figments of the imagination of persons who would sell unnecessary protection for an imaginary problem. The education and experience of the last decade have clearly demonstrated that this is not the case.

Virtually every user of electricity is affected by transients causing unexplained data losses and scrambled data on computers, excessive lighting replacements, excessive motor insulation deterioration, and much more. Heavy duty motors, sensitive medical equipment, personal computers, office equipment, and even household appliances are all equally subject to the damaging effects of transient activity. Transient activity can be divided into external sources and internal sources.

6.3.1.1.1 External Sources of Transient Activity

Power companies do everything possible to provide their customers with clean, steady power. Utility systems are designed to provide reliable bulk power. However, it is not feasible for them to provide continuous power of the quality required for a completely undisturbed computer operation. Also it should be noted that many pieces of equipment other than computers can benefit from disturbance-free electricity.

Power companies have no control over transients induced by lightning or high power switching at substation levels. Figure 6–1 illustrates the enormous energy content of a typical lightning waveform. Currents from a direct or indirect strike may enter conductors of a suspended cable or enter a buried cable by ground currents.

Either way, the surge will propagate in the form of a traveling wave bi-directionally on the cable from the point of origin. Severity of impact to the end user is directly proportional to the proximity of the lightning strike. If the facility is at a 10 to 20 pole distance from the strike, little harm will occur, since the surge current will have been dissipated by the utility ground system. Such is not the case

Protection Systems for Low Voltage, Low Power Systems 179

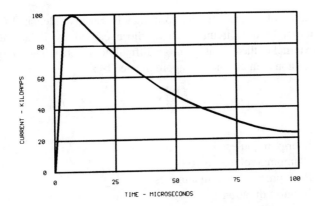

FIGURE 6-1 A typical lightning current waveform

where the strike is much closer. In this case, the residual current can migrate through the facility's service equipment and cause severe damage.

Other externally generated transients result from switching in nearby industrial complexes which can send transients back into the power line, causing damage to equipment.

6.3.1.1.2 Internal Sources of Transient Activity

Internally generated transients result from switching within the facility. Any time the flow of electricity is altered, such as in the simple act of turning a motor or light on and off, transient activity can result in an "inductive kick." An excellent example of this phenomenon can be seen in Figure 6–2. It shows the transients generated by turning off a fluorescent light.

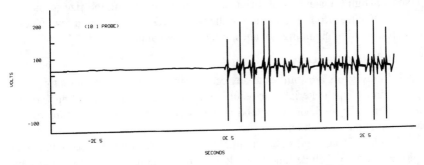

FIGURE 6-2 Transients generated by turning off a fluorescent light

6.3.1.1.3 Transient Energy

The energy of a transient waveform may be readily calculated for transients that are internal to the circuit, such as those caused by inductive switching. Transients external to the circuit are more difficult to quantify. The energy absorbed by the suppression element may be approximated by:

$$E = \int_0^\tau V_c I_p \, dt \tag{6.2}$$

Where
E = Energy in joules
V_c = Clamping voltage in volts
I_p = Peak pulse current in amperes
τ = Impulse duration

6.3.1.1.4 Practical TVSS and Performance Considerations

Technologies available today for transient overvoltage protection consist of metal oxide varistors (MOVs), gas discharge arrestors (commonly known as gas tubes), or solid state devices (TVS diodes and TVS thyristors). Each type of device is designed to serve a specific application.

Gas Tubes

Gas tubes are devices that employ an internal inert gas that ionizes and conducts during a transient event. Because the internal gas requires time to ionize, gas tubes can take several microseconds to turn on or "fire." In fact, the reaction time and firing voltage are dependent on the slope of the transient front. A circuit protected by a gas tube arrestor will typically see overshoot voltages ranging from a few hundred volts to several thousand volts.

Metal Oxide Varistors (MOVs)

MOVs are devices composed of ceramic-like material usually formed into a disk shape. High transient capability is achieved by increasing the size of the disc. Typical sizes range from 3 to 20 mm in diameter. MOVs turn on in a few nanoseconds and have high clamping voltages, ranging from approximately 30V to 1.5kv.

Solid State Devices

TVS thyristors are solid state devices constructed with four alternating layers of p-type and n-type material. The resulting device is capable of handling very high pulse currents. These devices respond in nanoseconds and have operating voltages that start at about 28V and up.

TVS diodes are solid state p-n junction devices. A large cross sectional area is employed for a TVS diode junction, allowing it to conduct high transient currents. These diodes respond almost instantaneously to transient events. Their clamping voltage ranges from a few volts to a few hundred volts, depending on the breakdown voltage of the device. The fast response time of the TVS diodes means that any voltage overshoot is primarily due to lead inductance.

Protection Systems for Low Voltage, Low Power Systems

Table 6–1 compares the characteristics of the most widely used TVS devices.

TABLE 6–1 Comparison of TVS Devices

Suppression Element	Advantages	Disadvantages	Expected Life
Gas Tube	• Very high current-handling capability • Low capacitance • High insulation resistance	• Very high firing voltage • Finite life cycle • Slow response times • Nonrestoring under DC	Limited
MOV	• High current-handling capability • Broad current spectrum • Broad voltage spectrum	• Gradual degradation • High clamping voltage • High capacitance	Degrades
TVS Diodes	• Low clamping voltage • Does not degrade • Broad voltage spectrum • Extremely fast response time	• Limited surge current rating • High capacitance for low voltage types	Long Limited
TVS Thyristors	• Does not degrade • Fast response time • High current handling capability	• Non restoring under DC • Narrow voltage range • Turn-off delay time	Long

TVS device protection levels may be divided into three categories namely: (i) primary protection, (ii) secondary protection, and (iii) board level protection.

Primary Protection

Primary protection is for power lines and data lines exposed to an outdoor environment, service entry, and AC distribution panels. Transient currents can range from tens to hundreds of kiloamps at these sources.

Secondary Protection

Secondary protection is for equipment inputs, including power from long branch circuits, internal data lines, PBX, wall sockets, and lines that have primary protection at a significant distance from the equipment. Transient voltages can exceed several kilovolts with transient currents ranging from several hundred to several thousand amps.

Board Level Protection

Board level protection is usually internal to the equipment; it is for protection against residual transient from earlier stages of protection, system generated transients, and Electro Static Discharge (ESD). Transients at this level range from tens of volts to several thousand volts with currents usually in the tens of amps.

6.3.1.1.5 Practical Surge Protection Circuits

MOVs, TVS diodes, and thyristors are available from many suppliers and have well established properties. The extremely fast clamping speed of TVS diodes may be combined with the high current handling capability of MOVs to provide optimum transient protection on AC lines. Figure 6–3 indicates four different kinds of TVSS circuits.

(a) Basic circuit of a multi-stage surge suppressor using MOVs, power zeners, and series inductors for power line surges
(b) Telecom line protection using gas discharge tube, MOVs, and zener diodes.
(c) Line driver/receiver protection
(d) Generic IC protection

Basic circuit of a multistage surge suppressor for power lines is shown in Figure 6–3 (a). For surges coming in on the line terminal, the current is shared by M_1 and M_2. The current is returned down the neutral and ground wires back to the AC breaker panel and building ground. Since the impedance of the wires for fast-rising spikes could be greater than that of the MOVs the wire impedance determines the current division. Since the neutral and ground wires back to the panel are typically the same size and length, current would normally be shared roughly equally. Thus, the 3000A test pulse may result in approximately 1500A pulse current into M_1 and M_2.

For short pulses, T_1 and T_2 do not absorb large amounts of the surge, since the L.dI/dt voltage developed across L_1 makes the first-stage MOVs absorb most of the current. The second-stage MOVs reduce transmitted surge voltage to well below the voltage across M_1.

Figure 6–3(b) indicates telecommunication protection elements and Figure 6–3(c) indicates the use of transients protection zener diodes such as Motorola 1.5 KE series being used for line driver/receiver protection. Figure 6–3(d) depicts the generic IC protection.

6.3.1.1.6 Transient Protection Standards

Committees such as ANSI, IEEE, and IEC have defined standards for transient waveshapes based on the threat environment. European or IEC transient standards include:

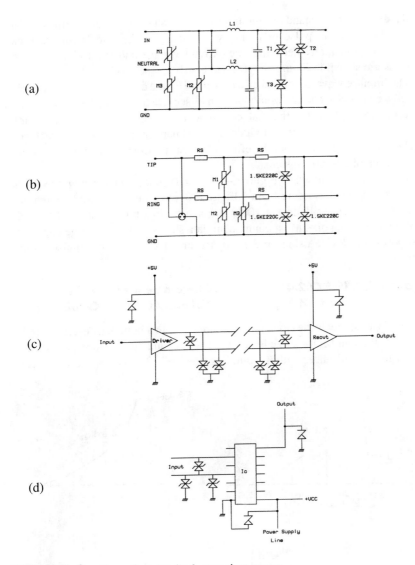

FIGURE 6-3 Protection circuits for transient surges

- IEC 1000-2 for Electro Static Discharge
- IEC 1000-4 for electrical fast transients (EFT)
- IEC 1000-5 for electrical transients

U.S. transient standards include:

- ANSI/IEEE C62.41-1991 for power line transients
- FCC Part 68 for telecommunication lines
- UL 1449, and various military standards

184 POWER ELECTRONICS DESIGN HANDBOOK

Regardless of the standard, the transient waveshape is usually defined as an exponentially decaying pulse or a ring wave (damped sinusoidal) for electrical surges. The exponentially decaying open circuit voltage and short circuit current waveforms are shown in Figures 6–4(a) and 6–4(b), respectively.

The impulse waveform is specified by its rise time and duration. For example, an 8 x 20µsec impulse current would have an 8µsec rise time from 10 percent of the peak current to 90 percent of the peak current and a 20µsec decay time as measured from the start of the virtual front to a decay to 50 percent of the value of peak current. The ring wave results from the effect of a fast rise time transient encountering the impedance of a wiring system.

The waveform shown in Figure 6–4(c) is a 0.5µsec 100kHz ring wave. The waveshape is defined as rising from 10% to 90% of its peak amplitude in 0.5µsec and decaying while oscillating at 100kHz. Each peak is 60 percent of the amplitude of the preceding one. Most component data sheets define the surge capability of a suppression device using an 8 x 20µsec or 10 x 1000µsec impulse current waveform.

6.3.1.1.7 IEEE C62.41-1991: "IEEE Recommended Practice on Surge Voltages in Low Voltage AC Power Circuits"

IEEE C62.41 identifies location categories within a building, described as A1, A2, A3, B1, B2, and B3 for surge locations. The "A" and "B" location prefixes represent wiring run distances within a building, the "1," "2," and "3," suffixes repre-

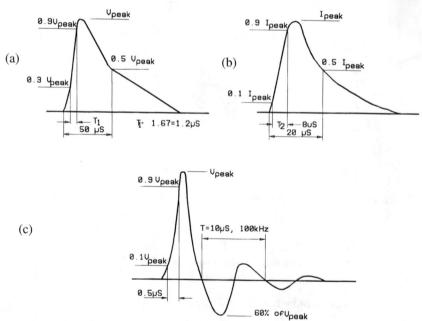

FIGURE 6–4 Surge waveforms (a) Open circuit voltage waveform (b) Discharge current waveform (c) Ring waveform

sent surge severity. "A" category locations receive their power after more than 60 feet of wiring run from the main power service entrance, with frequent exposure to comparatively low energy surges.

These low energy surges are created when internal equipment cycles on and off, collapsing the magnetic field around the lengthy building wiring carrying the current. The collapsing magnetic field induces surge voltage in the wires proportional to the current being carried by the wiring and the length of wiring.

External surges will be of a lesser threat in "A" locations than "B" locations due to the impedance protection provided by the inductance of the building wiring in "A" locations. IEEE specifies a low energy "ringwave" surge waveform for the "A" locations.

"B" category locations are within a building, close to the power service entrance, with greater exposure to infrequent, high energy surges originating outside the building. The "B" location category surges can be caused by lightning, power outages due to storms, and normal utility switching functions. A "combination wave" with high surge energy is specified by IEEE for these "B" locations.

"1," "2," and "3" denote low, medium, and high exposures respectively, in terms of the number and severity of surges, with "1" being the least severe and "3" being the most severe. A "B3" location, therefore, would have the highest exposure to surge energy, while a "A1" location would have the lowest incidence of surge energy.

TABLE 6–2 C62.41 Location Categories, Frequency of Occurrences and Surge Waveforms

IEEE Location Category	IEEE Exposure	2,000 Volt, 70 Ampere Ringwave Surges (.63 Joules)	4,000 Volt, 130 Ampere Ringwave Surges (2.34 Joules)	6,000 Volt, 200 Ampere Ringwave Surges (5.4 Joules)
A1	Low	0	0	0
A2	Medium	50	5	1
A3	High	1,000	300	90
		2,000 Volt, 1,000 Ampere Combination Wave Surges (9 Joules)	4,000 Volt, 2,000 Ampere Combination Wave Surges (36 Joules)	6,000 Volt, 3,000 Ampere Combination Wave Surges (81 Joules)
B1	Low	0	0	0
B2	Medium	50	5	1
B3	High	1,000	300	90

Table 6–2 shows possible annual surge magnitudes, frequency of occurrences, and surge waveform as extracted from the IEEE C62.41 Standard. Figure 6–5 depicts the building locations.

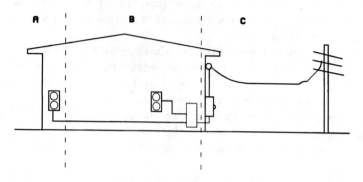

FIGURE 6-5 Building locations and C62.41 categories

6.3.1.1.8 UL 1449

Several years ago, Underwriters Laboratories established a uniform TVSS rating system by creating UL Standard 1449-1987. Although somewhat limited, it was the first step toward establishing benchmarks by which to compare TVSS products. Underwriters Laboratories established specific test criteria to determine the ability of a TVSS product to stop the travel of a transient voltage surge into protected equipment.

Underwriters Laboratories divided TVSSs into two categories: plug-in or cord-connected devices and hardwired or direct-connected devices. TVSS devices included in the plug-in and cord-connected category are subjected to a maximum transient surge impulse of 6000V with a short circuit current of 500A available. The voltage available is in a 1.2 by 50μs waveform and the current is available in an 8 by 20μs waveform. Devices included in the hardwired category are subjected to an impulse of 6000V with 3000A available in the same type of waveform (Lewis 1995). Underwriters Laboratories Inc. has recently proposed revised testing procedures for UL1449 (Hartford 1996).

6.3.1.1.9 Electrostatic Discharge and Circuit Protection

Recent years have seen various standards developed for testing the ESD capability of semiconductor products. These standards have been generated with regard to a specific need related to the electromagnetic compatibility of the system environment. They include the Human Body Model (HBM), Machine Model (MM) and the Charged Device Model (CDM). Each such standard relates to the nature of electrostatic discharge generated within a system application and the potential for damage to the IC. Among the better known standards are:

- Human Body Model using the MIL-STD-883, Method 3015.7
- Machine Model using EIAJIC121
- Human Body model using the IEC 1000-4-2 standard

Testing for ESD immunity is more broadly defined to include a device, equipment, or system. Both direct contact and air discharge methods of testing are used with four discrete steps in the severity level ranging up to 8KV and 15kV, respectively. In its simplest form, the Figure 6–6(a) test circuit provides a means of charging the 150pF capacitor, R_C, through the charge switch and discharging ESD pulses through the 330Ω resistor, R_D, and discharge switch to the Equipment or Device Under Test. The test equipment for the IEC 1000-4-2 standard is constructed to provide the equivalent of an actual human body ESD discharge and has the waveform shown in Figure 6–6(b).

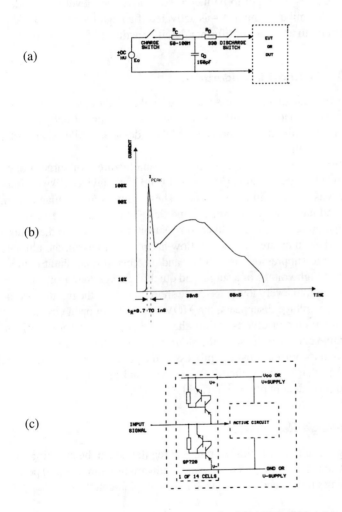

FIGURE 6–6 IEC 1000-4-2 testing (a) Simplified IEC 1000-4-2 test energy source (b) Typical waveform of the output current of the HBM ESD generator (c) SP 720 implementation for active circuit protection showing one out of 14 cells. (Copyright by Harris Corporation, reprinted with permission of Harris Corporation Semiconductor Sector)

Recently, semiconductor manufacturers such as Harris Semiconductors have introduced SCR diode arrays for semiconductor circuit protection. These devices cater for ESD protection requirements. The SP720, SP721, and SP723 are protection ICs with an array of SCR-diode bipolar structures for ESD and overvoltage protection of sensitive input circuits. They have two protection SCR-diode device structures per input. See Figure 6–6(c).

The SCR structures are designed for fast triggering at a threshold of one $+V_{BE}$ diode threshold above +V (positive supply terminal) or a $-V_{BE}$ diode threshold below V– (negative or ground). A clamp to V+ is activated at each protection input if a transient pulse causes the input to be increased to a voltage level greater than one V_{BE} above V+. A similar clamp to V– is activated if a negative pulse, one V_{BE} less than V–, is applied to an input. For further details Austin and Sheatler (1996) is suggested.

6.3.1.1.10 Practical Considerations

The market offers an almost endless variety of makes and models of transient-voltage surge suppressors. However, many manufacturers provide virtually no written specifications to describe the performance of their devices, while others offer exaggerated claims (Steinhoff 1991).

Most designers of electronic equipment, and manufacturers of surge protection devices, use metal-oxide varistors (MOV) and solid state devices (like bidirectional zener diodes) as voltage limiting devices. These devices are available from various suppliers, and have well established properties (Cohen and Mace 1991).

Ordinary surge protectors simply divert surges from the hot line to the neutral and ground wires, where they are assumed to flow harmlessly to earth, the ultimate surge sink. These surge suppressors use MOVs and/or other similar shunt components which sense the high voltage of a surge, and quickly change state from an open (non-conducting) circuit, to a very low impedance short circuit for the duration of the surge. When the surge voltage disappears, the MOV returns to an open circuit.

In this way, the protector diverts mainly the surge energy and not significant amounts of power line energy because of the short duration of the surge, unless the MOV has degraded in its normal wear out process to the point where it "clamps" on the power line. When that happens, the MOV explodes and fails, leaving the surge protector unable to provide any protection (Laidley 1991).

6.3.2 Voltage Regulators

In AC voltage regulators the limited range of regulation can be anything from less than 1 percent change, over a given line, and load variation range, to 10 percent or greater, depending on the need and the technology implemented.

6.3.2.1 Motor Driven Variacs

The oldest and most simple type of automatic voltage regulators are the motor driven variacs, where the tap setting of a transformer is changed with the feed back signal from an output monitoring circuit. Even though these are used in many sites they are very slow, bulky, and sometimes have reliability problems.

6.3.2.2 Electronic Tap Changers

Another more practicable type with light weight and low cost are the electronic tap change systems. Electronic tap changers are electronically controlled devices designed to compensate for line-voltage fluctuations by sensing and switching buck or boost taps to vary the primary-to-secondary turns ratio of a power transformer (see Figure 6–7). Tap-changers are more efficient (low 90 percent range), cooler, and quieter than ferroresonant Constant Voltage Transformers (CVTs), with better output stability. However, they lack the ruggedness of the CVT, and are accordingly less often considered as an industrial alternative.

There are two main types of tap-changers. The more popular design incorporates an isolation transformer. It also provides good common-mode noise attenuation but suffers high output impedance. The less popular is based on the auto transformer, which provides low output impedance but lacks in isolation and noise protection. Under certain conditions, both types can exhibit a response known as "tap dancing," in which input oscillates between taps, continually "notching the line," and creating unstable voltages that can damage SCRs and other control electronics.

Electronic tap-changers generally do not recover well from line notches. Some microprocessor-controlled units drop their output for several line cycles after a half-cycle interruption. In some designs, a line drop below the tap-changer's low-voltage threshold (which can occur during a severe brownout) can cause the sensing/switching electronics to shut the regulator off as a self-protective measure. In either case, the tap-changer in effect creates its own blackout, allowing its load to crash. One other problem inherent in electronic tap-changers is that their electronic sensing and control systems are necessarily exposed to raw line power. Therefore, they are vulnerable to the same line noise and transients they are supposed to guard against.

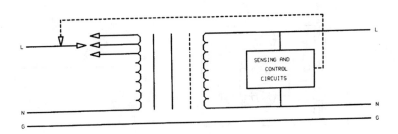

FIGURE 6–7 Block diagram of an electronic tap changer

6.3.2.3 Thyristor Driven AC Regulators

SCR controlled AC regulators are also a low cost light weight implementation. The most common form of thyristor circuit for AC voltage control is shown in Figure 6–8(a) (the thyristors could be replaced by a triac). This has been widely used in applications such as lamp dimmers, where a wide control range is required for loads tolerant to high levels of distortion.

The simple phase control circuit of Figure 6–8(a) will not meet the requirement of producing a nominal output level. This will require the introduction of an auto transformer as shown in Figure 6–8(b). For an rms output voltage equal to the nominal supply voltage (rms value of V_o) the thyristor firings must be delayed by about 95° when the supply is at its maximum ($1.2V_o$). This delay reduces to zero if the supply is reduced to its lower limit ($0.8V_o$). There could be many variations to this basic implementation in the practical voltage regulators.

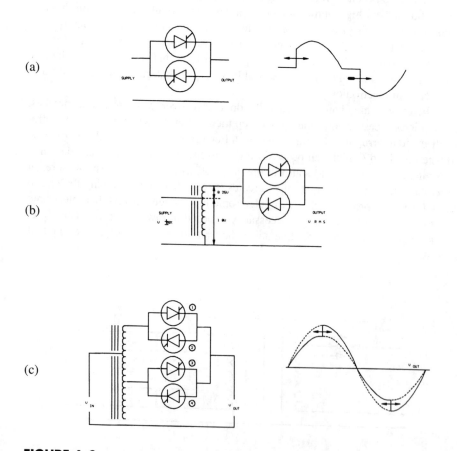

FIGURE 6–8 Thyristor control (a) Simple thyristor phase control and output waveform (b) Single tap regulator (c) Two tap regulator and output waveform

This form of regulator introduces considerable distortion in the output voltage. Under the worst condition, when the supply is at $1.2V_o$ and thyristor firing is held back by about 95°, the total distortion is over 50 percent. Even if the regulation range of the regulator were reduced to ± 5 percent the distortion would still be over 20 percent.

The use of two taps on the auto transformer as shown in Figure 6–8(c) results in very significant improvements in regulator performance. Its operation when feeding resistive loads is simple. At the start of each half cycle one of the lower thyristors conducts, supplying voltage to the load. Some time later in the half cycle, one of the upper thyristors is triggered. This increases the voltage on the load and by so doing commutates the lower thyristor. At the end of the half cycle, the upper thyristor is extinguished.

The output voltage can be continuously varied between the two values available at the transformer taps by altering the time of triggering the upper thyristors. The distortion is considerably lower than the single tap circuit, being zero at both extremes of input voltage and about 15 percent for the worst condition. There will be proportionally lower distortion if the regulation range is reduced below ± 20 percent.

Thyristor regulators when used on inductive loads could cause special transient situations, which may stress the thyristors. This type of situations and the inherent waveform distortion on basic thyristor regulators make these less popular than some other types, unless these are further buffered by special output filters, etc.

6.3.2.4 Ferroresonant Regulators

Ferroresonant transformer or constant voltage transformer (CVT) has been commercially available for over four decades.

(a) Conventional Transformer
(b) Flux vs. Magnetizing current
(c) Ferroresonant transformer–basic arrangement
(d) Ferroresonant transformer with a harmonic neutralizing coil

The easiest way to describe a ferroresonant transformer's operation is to compare it with an ordinary power transformer (Figure 6–9 (a)). If a typical transformer operates at a high enough primary current, its core becomes saturated with flux lines. Increasing the primary current will not then increase the secondary voltage.

Consequently, operating an ordinary power transformer in saturation will create a voltage regulator, though such operation is impractical because primary saturation current is very close to a short circuit. Normal current in the I_1-I_2 region shown in Figure 6–9(b) causes a linear change in flux, but once the current reaches saturation region (I_4-I_5), the resulting flux changes are small. To the left of the point I_3, flux changes are in direct proportion to primary current: to the right of the I_3, the curve becomes non-linear and changes in current have less effect on flux.

The ferroresonant transformer, on the other hand, is designed to operate in saturation (Figure 6–9(c)). However, since it is the secondary coil that operates in saturation, not the primary, more flux must be produced in the secondary core leg than

192 POWER ELECTRONICS DESIGN HANDBOOK

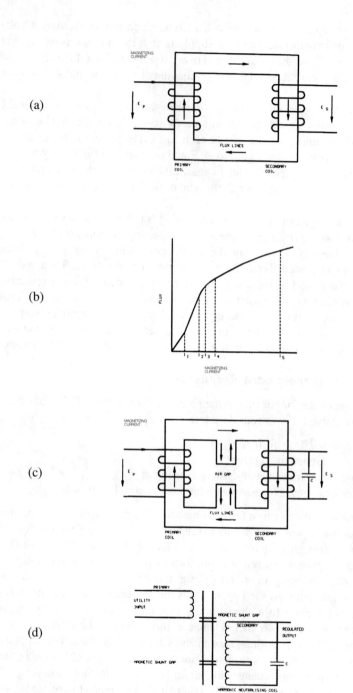

FIGURE 6-9 Comparison of ferroresonant transformer and conventional transformer

in the primary. The core structure is modified by the addition of a magnetic shunt to loosen the coupling between primary and secondary. Because of the air gap in the shunt, the core section has a high reluctance and the main path for magnetizing flux is through the outer core. As the input voltage increases, the magnetizing flux through the core section increases, and with it, the inductance of the secondary winding. The addition of an output capacitor alters this.

Once the reactance of the winding equals that of the capacitor, the two resonate, producing a higher output than the turns ratio voltage alone. This effect is similar to a series resonant circuit where point of resonance increases the voltage across the capacitor substantially above the impressed voltage.

The output waveform of a basic ferroresonant transformer will be like a square wave (due to flattened top) and is suitable for many loads, but may not satisfactorily provide power for some electronic loads without the addition of a harmonic filter choke (Figure 6–9(d)). With this filter, a large part of the output harmonic content is canceled producing a sine wave voltage with reasonably low total harmonic content.

A properly designed conditioner could regulate the output voltage to ± 2 percent or ± 3 percent. With the input varying as much as ± 20 percent, the output total harmonic distortion should be no greater than 5 percent or 6 percent.

One of the advantages of ferroresonant transformers is their ability to attenuate normal mode noise voltage transients. This attenuation is up to 10 times more than conventional tap changer regulators provide. Since the secondary winding operates in saturation, transients and spikes are clipped. A ferroresonant transformer has some very desirable overload characteristics. If the output is short circuited, the current increases by only about 80%. A well designed ferroresonant transformer can maintain a short circuit indefinitely.

Note that the ferroresonant transformer is a tuned circuit, which is frequency sensitive. Typically, a 1 percent change in frequency produces a 1.5 percent change in output voltage. Another draw back of ferroresonant transformers are their low efficiency, particularly at the no load and low load conditions.

Figure 6–10(a) describes the operating characteristics typical for a ferroresonant line regulator for input voltage versus load power factor. Figure 6–10 (b) shows the output voltage versus percent load. These curves are not intended to describe the technology in total capability, but rather are presented to show the typical performance of a particular design.

6.3.2.5 Other Types of Voltage Regulators

Within the past several years, new solid state techniques were proposed for AC line voltage regulation, as well as power conditioning. Some of these techniques are based on bipolar power transistor arrays with simple buck-boost transformers as shown on Figure 6–11.

Simplified block diagram of such an AC voltage regulator is shown in Figure 6–12. Basic techniques used for the conditioner could be explained using Figures 6–11(a) and (b). Referring to the block diagram in Figure 6–11(a) primary (a)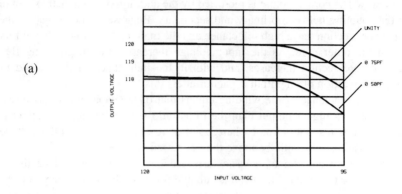

(b)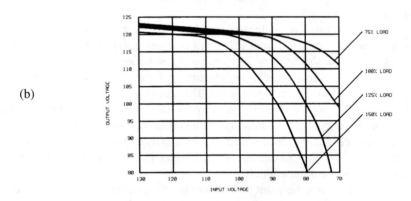

FIGURE 6-10 Characteristics of ferroresonant regulator (a) Output voltage vs. power factor for a ferroresonant regulator (b) Output voltage vs percentage load for a ferroresonant regulator

of a buck-boost transformer T1 is connected in series with a bridge rectifier and an impedance Z. Impedance Z created by a multi-element bipolar power transistor array as per Figure 6–11(b) is controlled by a feedback control circuit which is designed using low voltage solid state circuitry. Bridge avoids voltage reversals across the bipolar power transistor array. As per configuration shown in Figure 6–11(b) the effective impedance across the transistor array is dependent primarily on the components R_b and R_e (Kularatna 1988, 1990, and 1993). In the practical implementation of the technique R_{e1} to R_{em} could be implemented with transistor elements of a multistage opto isolator. Figure 6–11(c) shows the required impedance across the array V_s input voltage and output load current.

The low voltage components used for feedback control and other functions are placed in isolation from the line voltage handling blocks and this improves the reliability of the design. Further, this technique does not have sensitivity for line frequency variations or power factor etc. and waveform fidelity (of the basic imple-

(a)
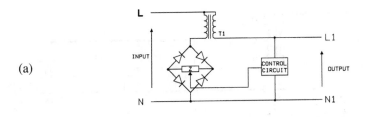

(b)
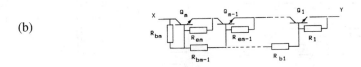

(c)
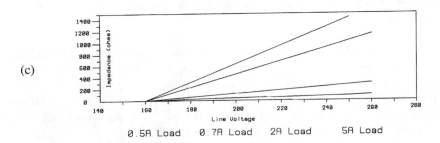

FIGURE 6-11 BJT array based AC line regulation (a) Basic technique (b) Arrangement of the multi-element transistor array (c) Impedance of transistor array vs line voltage and load current

mentation itself) is far superior to ferroresonant systems. Also the technique could be easily modified to provide harmonic control, where transformer design itself is truly simple compared to a ferroresonant transformer. This technique is easily applicable for VA ratings from 200VA to 3KVA and can be easily mixed with electronic tap changers (for efficiency improvement purposes rather than regulating purposes) as discussed in Kularatna (1988, 1990, and 1993).

Another new approach to AC line voltage regulation by switched mode techniques is described in Zhang (1994) and Bhavaraju and Enjeti (1993).

6.3.3 Isolation Transformers

Isolation transformers are specially designed transformers that attempt to limit the coupling of noise and transient energy from the primary side to the secondary side by "isolating," or physically separating those two windings, and adding electrostatic shielding bonded to the ground wire (see Figure 6–13).

196 POWER ELECTRONICS DESIGN HANDBOOK

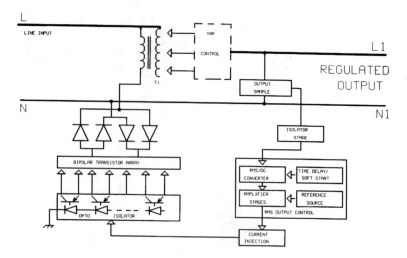

FIGURE 6-12 Block diagram of a practical voltage regulator based on bipolar power transistors

As seen in Figure 6–14(a) common-mode noise is generated between either of the power lines and ground. The capacitor shown between the primary and secondary coils is present in all transformers and is called coupling capacitance. The presence of coupling capacitance provides an AC path for common-mode noise to pass directly to the secondary coil and the load. This coupling capacitance can be reduced by the addition of shielding between the coils.

In Figure 6–14(b) a simple Faraday shield has been added between the primary and secondary windings. As a result, C is greatly reduced, and C_1 shunts current directly to ground through the shield. This attenuates current through C_2 and the secondary winding, and can reduce common-mode voltage effects in the secondary by as much as 50 dB, when compared with the unshielded transformer.

Common-mode transients of greater than 2000 V can be expected on any power line. With a ratio of 50dB, 2000-V transient would appear as 6.3 V on the sec-

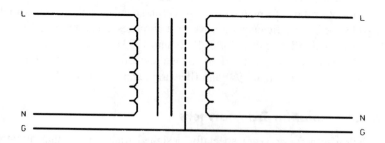

FIGURE 6-13 Isolation transformer

(a)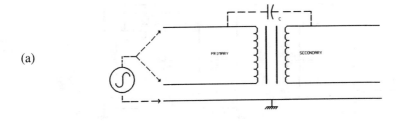

(b)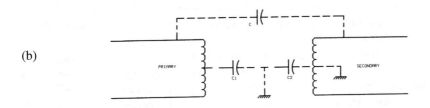

FIGURE 6-14 Effect of coupling capacitance (a) Capacitor C providing a direct path for common-mode noise (b) Reduction of C due to increase of effective distance between primary and secondary

ondary. This degree of isolation may not be adequate for many installations. In some cases it is desirable to achieve 100 dB or more input to output isolation.

The shielding method shown in Figure 6–15 is capable of achieving common-mode isolation exceeding 140 dB (or 10,000,000: 1). By enclosing both the primary and secondary windings in box shields, and including a Faraday shield between the two, maximum isolation can be achieved. To obtain the best effect, the primary shield must return to the utility ground, the Faraday shield connects to a separate earth ground, and the secondary shield connects to the load ground.

If this sort of isolation is really necessary, consideration must be given to the fact that there is now no direct path between the utility ground and the load ground. Safety must be considered at this point, and consideration given to what is given up in isolation in order to ensure the integrity of the ground system.

Normal mode noise also falls into about the same category as surges described earlier. In attempting to attenuate surge voltage, the designer's goal is to design the transformer to distinguish between what is power and what is surge or noise. The key to doing this is the difference in the frequencies. The power signal is constantly low in frequency, say, 60 or 50 Hz, while surges and noise will appear at much greater frequencies.

The transformer by itself can do nothing to attenuate slowly changing voltages applied to its input, and a voltage surge that increases the line voltage by 15 percent,

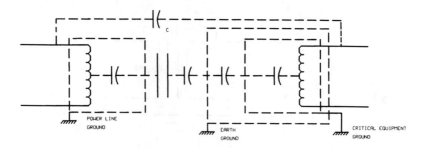

FIGURE 6-15 Isolation transformer with two box shield

and stays there for a long period of time, will be passed right on through the transformer. (See Figure 6–16). But the damaging surges discussed earlier are all of the type that have fast rising and leading edges, and are over within a relatively short period of time.

If a transformer could be intentionally designed to have a high leakage inductance (similar to ferroresonant transformer), as the frequency increases on the input to the transformer, its ability to pass voltage to the secondary is diminished. The result is that a transformer intentionally designed to maximize the leakage inductance will pass line frequency voltage well, but any high frequency voltage riding on top of the line voltage will be attenuated, just as if it were passing through a low pass filter.

Figure 6–15 shows how a typical isolation transformer, designed to maximize leakage inductance within reason, will perform. The addition of output capacitance will greatly enhance the performance, as shown.

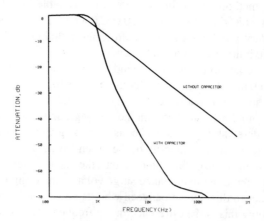

FIGURE 6-16 Transverse mode attenuation curve

6.3.4 Line Conditioners

Line conditioners, also known as power line conditioners, provide multiple types of protection in one device. These systems offer characteristics of two or more enhancement devices and usually cost less than the combination of the individual power enhancers. For example, a TVSS and an isolation transformer with electronic tap changer may be considered as a power conditioner in practice.

The most common power conditioner is the ferroresonant regulator. These are designed to resonate with a suitable capacitor and the stored energy within the resonating system (system inductance and capacitor) provides some back-up for momentary outages, usually less than a fraction of a cycle to few cycles of the AC line. However low efficiency, high weight and volume of these "ferros" do promote considering other types of power conditioners.

6.4 Power Synthesis Equipment

Power synthesis equipment uses the incoming power as merely an energy source, creating "new power" for the critical load, which is completely isolated from the incoming waveform. Common types are stand by power sources (SPS), motor generators (MG) and uninterruptible power supplies (UPS). The next chapter details these systems.

References

1. Steinhoff, H. "Transient–Voltage Protection in the Real World." Power Quality Proceedings, September 1991, pp 32–38.
2. Cohen, R. L. and K. Mace. "Lifetime and Effectiveness of MOV–Based Surge Protectors." Power Quality Proceedings, September 1991, pp 1–7.
3. Laidley, W.H. "No Nonsense Computer Surge Protection: Wrong Choices Can Cause Failures." Power Quality Proceedings, September 1991, pp 21–24.
4. Kularatna, A. D. V. N. "Low Cost, Fast AC Regulator Provides High Waveform Fidelity." PCIM USA, May 1988, pp 76–78.
5. Kularatna, A. D. V. N. "Low Cost Light Weight AC Regulators Employing Bipolar Power Transistors." Power Conversion International Proceedings—USA, October 1990, pp 67–76.
6. Kularatna, A. D. V. N. "Techniques Based on Bipolar Power Transistor Arrays for Regulation of AC Line Voltage." EPE Proceedings, Vol. 7, September 1993, pp 96–100.
7. Zhang, X. Z. "Analysis and design of switched Mode AC/AC voltage regulator with series compensation." Power electronics and variable speed drives–PEVD-94, October 1994, pp 181–187.
8. Bhavaraju, V. B. and P. Enjeti. "A fast acting line conditioner for sensitive loads corrects voltage sags." Power Quality Proceedings, USA, October 1993, pp 343–350.
9. Lewis, P. "Transient Voltage Surge Suppression Response Time." Power Quality Assurance, September/October 1995, pp 56–63.
10. Harford, J. R. "What Changes to UL 1449 Standard for Safety Transient Voltage Suppressors May Mean to You." Power Quality Assurance, January/February 1996, pp 48–51.
11. Austin, W. and R. Sheatler. "Transients vs. Electronic Circuits: Survival of the fittest, the need for transient protection (Part II)—ESD Immunity and Transient current capability for electronic protection array circuits." PCIM, May 1996, pp 46–53.

Bibliography

1. Harold, P. "Units Power PCs in Rough Times." *EDN*, 25 May 1989, pp 73–84.
2. Michele, F. "Stabilizing the Mains with Ferro-resonant Technology." *Electronics & Wireless World*, Feb. 1992, pp162–163.
3. Nay, J. "Shielding Improves Isolation Transformers Noise Rejection." *PCIM,* October 1989, pp 60–64.
4. Intertec International. "Power Quality Sourcebook." 1991.
5. Wilfong, H. D. "Selecting Transient Suppressors." Power Quality/ASD Proceedings. October 1990, pp117–122.
6. Goulet, K. "Automation Equipment of the 90's—Power Conditioning Equipment of the 60's." Power Quality/ASD Proceedings, October 1990, pp 64–77.
7. Laidley, W. H. "Computer Network Power Protection: Problems, Myths and Solutions." Power Quality/ASD Proceedings, October 1990, pp181–191.
8. Thompson, R. "AC Voltage Regulators." *Wireless World*, July 1973, pp 339–344.
9. Hart, H. P. and R. J. Kakalec. "The derivation and application of design equations for ferroresonant voltage regulators and regulated rectifiers." IEEE transactions on Magnetics, March, 1971, pp 205–211.
10. Hasse, P. "Overvoltage Protection of Low Voltage Systems." Peter Perigrinus Ltd., 1992.
11. Clark, J. W. *AC Power Conditioners*. Academic Press, Inc., 1990.
12. IEEE Standard C 62. "Guides and Standard for Surge Protection." 1990.
13. Russell, W. "Transients vs. Electronic Circuits: Survival of the Fittest, The Need for Transient Protection (Part I)." *PCIM*, April 1996, pp 66–71.
14. Maxwell, J., N. Chan, and A. Tempelton. "Transients vs. Electronic Circuits—Part III: Multilayer Varistors." *PCIM*, June 1996, pp 62–70.
15. Armstrong, T. "Surge Protection for Mobile Communications." *Electronic Design*, November 4, 1996, pp 133–141.
16. Lurch, H.S. "Impact of Protecting Facilities to the Revised IEEE C62.41 Standard." *Power Quality*, April/June 1992, pp 72–73.

CHAPTER 7

Uninterruptible Power Supplies

7.1 Introduction

With the introduction of mainframe computers in the 1950s, power system engineers had to take a hard look at energy needs from a quality angle. As systems grew in complexity, it became more apparent to the engineers that the power that kept the systems running was raising havoc, causing equipment failures and corrupting data. As a result, the Uninterruptible Power Supply (UPS)—a back-up to stand between the commercial power supply and the computer—was born. The first uninterruptible power supply equipment, then known as no break power supplies, were of rotary design as shown in Figure 7–1. The market at that time for these systems were in mainframe computers, communications, and radar, etc.

During the last two decades there has been a shift from mainframe computer system sales to minis, micros, and portable computers and more recently towards network systems. With this shift, several changes have occurred in the UPS industry. The three decades from 1960 gave birth and evolution to different kinds of static type UPS systems where no rotating electrical machinery was used as major system components.

The physical size of UPS systems has decreased dramatically. Also, the cost has been reduced from over $2 to $1 per watt. As the world has migrated from the mainframe to the client/server distributed networking environment, the UPS market has shifted from the large multi-module UPS systems to the small single phase UPS systems. As per U.S. industry estimates (*PQ Assurance Journal*), in 1992 there was a total of US$ 1.225 billion worth of sales which had a breakdown of US$ 964 million commercial, US$ 245 million industrial and US$ 16 million residential components. In 1997 these figures were US$ 2.6, 2.1, 0.5 and .027 billion respectively

202 POWER ELECTRONICS DESIGN HANDBOOK

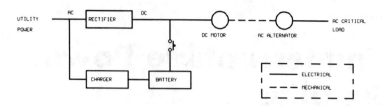

FIGURE 7-1   An old type of a rotary UPS System

for total, commercial, industrial and residential components. In the year 2002 total sales are expected to rise up to 3.94 billion US$ with 3.17 billion commercial and 732 million of industrial components. Industry estimates further indicate that over 71 percent of the UPS sales are for units less than 30 kVA ratings while 24 percent account for units of the capacity between 31 to 500 kVA ratings. Very large systems with ratings over 500kVA account for the rest.

In 1992, out of total UPS market, only 37 percent was made up of large three phase systems. The 63 percent balance was made up of small single phase UPS units. It was anticipated that by 1997, the UPS market to have large three phase modules will only represent 25 percent of that total with the balance of approximately 75 percent being comprised of small single phase units (Katzaro,1993). This chapter provides an overview of UPS systems with particular attention to single phase low power UPS systems used in modern information environments.

7.2   Different Types of Uninterrupted Power Supplies

Present day UPS systems can be divided in to three basic topologies, namely: off-line, hybrid, and on-line types. Each topology, in turn, can feature one or more technical variations, although the basic operation is about the same within each group. All UPS use an internal battery that produces AC power via an inverter. How and when this inverter comes into play largely determines the effectiveness of the UPS.

7.2.1   Off-Line UPS

Off-line UPS systems are the simplest forms of back up power systems. A block diagram of an off-line system is shown in Figure 7-2(a). The off-line type

UPS systems normally operate off-line and the load is normally powered by the utility line. When the utility power excursions are of such magnitude that they are beyond acceptable limits, or fail altogether, the load is transferred from the utility line to the UPS. The actual transfer time is usually very fast, in the sub-cycle range, however, the detection time may be longer and therefore, off-line UPS may not be as reliable as an on-line system.

The major advantage of the off-line UPS systems is lower cost, smaller size and weight and higher efficiency, since most of the time the UPS system is off-line and the load is powered by the utility. However, the disadvantages of the off-line UPS is that switching to the inverter is required when the load is most vulnerable, i.e., upon failure of the normal power source.

Here, the term UPS is really a misnomer because the inverter is normally off. For this reason off-line UPS are also known as standby power sources, or SPS.

Figure 7–2(b) indicates the general arrangement of an off-line UPS system. When the line voltage is within acceptable limits, the load is powered from the input utility supply. During this operation a rectifier block keeps the battery bank charged. When the input voltage sensor block detects an out-of range input voltage, a relay disconnects the incoming supply and load gets connected to the inverter block.

The only significant advantage of off-line the UPS is low cost. This is possible because the inverter in these systems is normally off so the charging and sense circuits are simple and inexpensive. These units provide no line conditioning or voltage regulation and provide only limited surge and spike protection. During sustained low voltage periods (brownouts) an SPS can inaccurately detect a blackout and prematurely switch onto battery.

If a site experiences sustained brownouts or takes successive low voltage hits, an SPS can completely discharge its battery and "crash" the system. In addition, an SPS switching time increases as the utility voltage decreases. It is not uncommon for a unit with a 5msec transfer time at 120VAC to exceed 15msec at 100 VAC. Because a brief period of low voltage precedes most blackouts, this may place the system at even great risk. Figure 7–2(c) shows the typical oscillograph of a off-line UPS at the transfer.

7.2.2 Hybrid UPS

These units are nearly similar to the off-line UPS units, but with the addition of a ferroresonant or electronic line conditioner that provides voltage regulation and energy storage (using a resonant circuit) in an attempt to ride through the glitch caused by switching on to the battery. A hybrid UPS with an electronic line conditioner is shown in Figure 7–3(a).

204 POWER ELECTRONICS DESIGN HANDBOOK

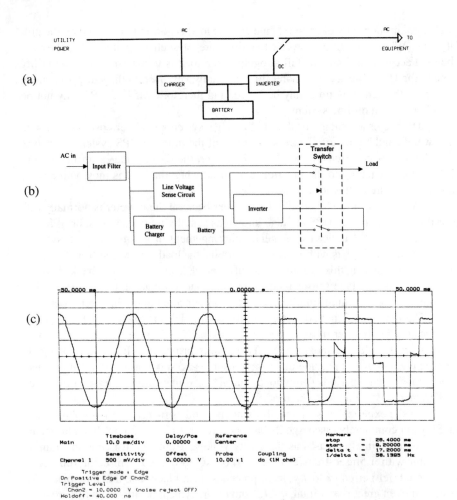

FIGURE 7-2 Off-Line UPS (a) Simplified block diagram of an off-line UPS (b) Block diagram of an off-line UPS showing the transfer process (c) An oscillograph of AC output during the transfer process of a typical off-line UPS

A hybrid UPS topology (sometimes called "Triport," "Line Interactive," "Utility Interactive," "Electronic Flywheel," "Hot Standby" "Bi directions," or "No Break") with a ferroresonant line conditioner is shown in Figure 7-3(b).

Some hybrid UPS systems eliminate or minimize the switching glitch by discharging the storage charge retained on the capacitors on the electronic or ferroresonant conditioner to bleed onto the load line while the unit switches to its battery.

In reality, this "ride-through" concept does not always work. During this switchover the output voltage can drop as much as 35V for a 120V nominal output system. Like SPS, hybrids can misinterpret brownouts as blackouts and prematurely switch to their batteries. Also like SPS, switching time can increase during low volt-

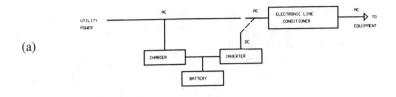

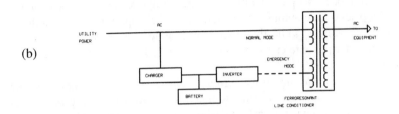

FIGURE 7-3 Hybrid UPS (a) A hybrid UPS with an electronic line conditioner (b) A hybrid UPS with ferroresonant line conditioner

age conditions, often exceeding the ride-through of the conditioner, as well as the hold up capability of a switching regulator load.

7.2.2.1 Line Interactive UPS Systems

Bell Labs in the mid 1970s proposed a new technique for UPS applications as shown in Figure 7–4(a). This topology, based on an originally patented technique in 1968, referred to as a "triport," uses a ferroresonant transformer with three power ports, two AC input ports (line and inverter) and the AC output port (Rando 1978).

The first triport was implemented with an impulse commutated SCR inverter (Rando 1978) which was operational at all times to safeguard against commutation failure at the instant of AC line failure. This guaranteed an operational inverter when battery power was needed to support the critical load. Even though the inverter was always on, its output was adjusted in voltage amplitude and phase to match the AC output. This way, all output power was drawn from the AC input. Since power is flowing into only one port of the triport, it is classified as a single power path line-interactive UPS.

Note that the Bell Labs topology was capable of dual power path operation. This was called the "share" mode (Rando 1978) and occurred when the inverter phase was shifted with respect to the AC output. In this mode, AC input and battery power was fed to the output. This mode was only used when transferring from battery power to line power where single power path operation ensued.

With the power semiconductors such as MOSFETS and IGBTs entering the market line interactive UPS system designers were able to replace the SCRs with the newer devices. Early design advancements were to replace the SCRs with power MOSFETS as the IGBTs were expensive.

Many variations to the basic technique discussed above are used in practical systems available in the market as line interactive models. Figure 7–4(b) depicts a single phase line interactive UPS with a bi-directional converter. The UPS consists of a three port transformer (triport), a bi-directional converter, two chokes, and a battery.

The bi-directional converter regulates the output voltage when the utility is present, provides output power when the utility fails, and also serves as a battery charger. The converter is large enough to deliver both the rated output power and the battery recharge power (typically 10 percent of maximum output power). The bi-directional converter is a pulse width modulated bridge regulated by two control loops. One loop senses the output voltage and varies the amplitude (pulse width) of the converter AC voltage while the other loop monitors the battery voltage and adjusts the converter phase to charge the battery. Figure 7–4(c) is an equivalent circuit of the line interactive UPS.

It can be shown that the utility current could lag or lead depending on the value of the utility voltage (Handler and Rangaswamy 1989) as per Figures 7–4(d) and 7–4(e). For further details on triport design Gupta and Handler (1990), Rando (1978), and Handler and Rangaswamy (1989) are suggested.

The triport design was very popular, however it resulted in a UPS that was heavy, noisy and had only limited range for correction for input voltage variations. Considering this drawback in the triport design some manufacturers such as Liebert Corporation have introduced designs based on auto-transformers. The auto-transformer could be designed with taps that allow correction of mains voltage over a wide range (Figure 7–5(a)). The inverter could drive a winding on the auto-transformer and provide backup when the mains was beyond control. In these designs leakage inductance of the auto-transformer was usefully utilized as an energy storing inductor for the charging mode. By using complex algorithms in the microcontroller subsystem, the inverter could be configured to run "backwards" and charge the batteries. However the design becomes bit tricky as the transformer needs to be designed with a well controlled built in leakage inductance which could store the

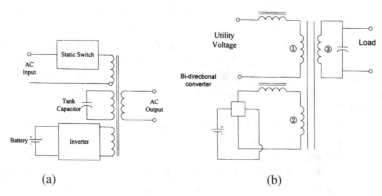

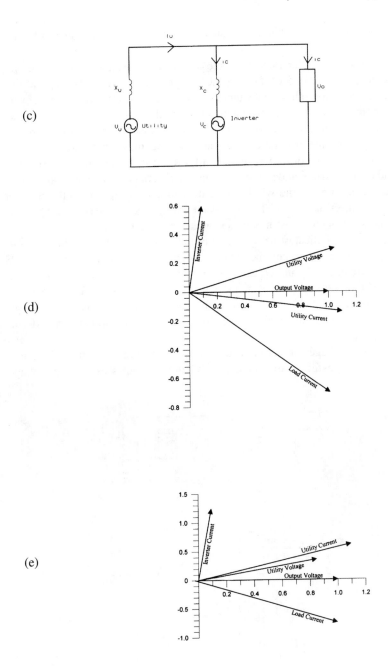

FIGURE 7-4 Triport UPS system (a) Basic block diagram (b) Design using a bi-directional converter (c) Equivalent circuit of line interactive UPS (d) Phasor diagram when utility voltage is higher than nominal (e) Phasor diagram when utility voltage is lower than nominal

proper amount of energy to charge the batteries. The concept is indicated in Figure 7–5(b).

A simplified block diagram of an auto-transformer based line interactive system is shown in Figure 7–5(c). During the charger mode two FETs, Q_4 and Q_3, switch on, charging the leakage inductance of the auto transformer. In the next cycle Q_2 and the body diode of the Q_3 switch on, charging the battery. In this case the circuit operates as a two quadrant boost converter. In this case 50Hz FETs are not switched as part of the battery charger control. The microcontroller subsystem which runs a complex algorithm to generate FET drive signals, monitors the battery voltage and charge current and adjusts the converter's conduction cycle maintaining the battery's charging regime within prescribed limits. Compared to the inverter mode where connections T_1 and T_2 serve as the transformer primary winding, these terminals are seen as the transformer's secondary in the charging mode.

Figure 7–5(d) indicates the operation of the inverter mode. The power semis Q_1 and Q_3 are switched by complementary 50 Hz signals which control the UPS base output frequency. Q_2 and Q_4 are switched by a high frequency signal such as 20 kHz PWM waveform to control the sine wave UPS output. This process is explained in section 7.3.2.3 as the same technique is used in the on line UPS systems with sine wave output. Figure 7–5(e) indicates the waveform at output during the transfer process.

If the input voltage varies and the transformer output remains in specifications there will be no transfer to inverter mode. When the input sees an over voltage,

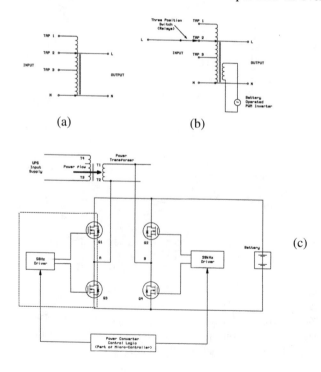

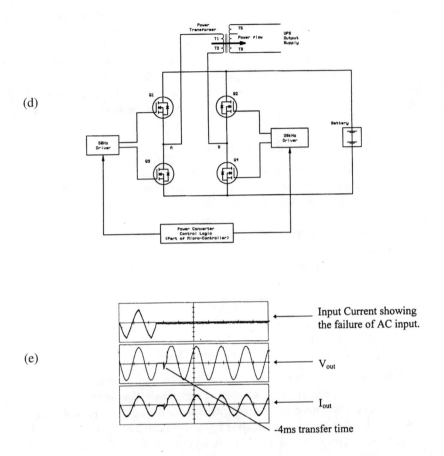

FIGURE 7-5 Line interactive UPS system with an auto-transformer (Courtesy of Liebert Corporation, USA) (a) Basic auto-transformer (b) Auto transformer with battery operated PWM inverter (c) Operation in charger mode (d) Operation in inverter mode (e) Output during the transfer process—a typical waveform

surge or a spike, associated circuits detect this and transfer the load to battery. When the disturbance is over it will transfer back to mains. If this happens quite often, the UPS may switch back and forth between mains and battery. If this type of mains problem is common, a line interactive UPS may not be the best solution. In this situation an on-line UPS may be required.

A line-interactive UPS will provide all the features of off-line models, with the addition of a "boost" and "buck" feature, which allows the load to be supplied from the mains by providing a low and high voltage mode of operation. Without this feature, the UPS would often be running on battery, and so when a full power failure occurred, the battery would be partially discharged, and the autonomy time may be insufficient for an orderly shutdown. Line-interactive UPS are suited to situations where the supply is prone to sags and surges. Line interactive and off line UPS systems normally draw high crest factor currents from the utility due to the fact that the load is either connected directly to the mains (off line) or connected to the mains through a transformer (line interactive). On line UPS systems on the other hand because of a power factor corrected AC to DC converter, have the capability to draw undistorted (low crest factor) currents which minimize the stress to the power grid.

7.2.3 On-Line UPS Systems

On-line UPS systems as shown in Figure 7–6 are not subject to many basic problems introduced by the previous types. Since they continuously regenerate clean AC power, they provide the highest level of protection available, regardless of utility line condition. A well designed on-line UPS protects against blackouts, surges, sags, spikes, transients, noise, and brownouts. Until recently their one drawback has been a higher price.

These systems have the following useful characteristics:

a) No switching involved
b) 100 percent line conditioning and regulation
c) Good sustained brownout protection
d) Typically sinusoidal output
e) Power factor correction and higher reliability

However, these units have much more complex designs than off-line or hybrid types, while the price, weight, and volume are higher. The inverter of an on-line UPS supplies continuous power to the critical load. Under conditions of overload or loads with high inrush currents which is beyond the capacity of the inverter, the static bypass switch provides mains power to the load. Basic components of an on-line UPS are shown in Figure 7–7.

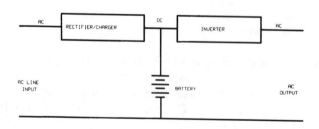

FIGURE 7–6   An on-line UPS

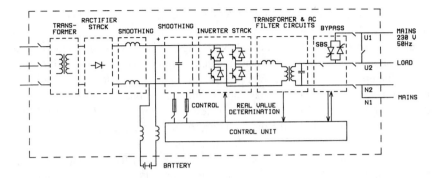

FIGURE 7-7 An on-line UPS system with static bypass, input isolation transformer, and other blocks

7.3 UPS System Components

Solid state UPS systems are comprised of several major elements. Those are the (i) rectifier and battery charger, (ii) inverter, (iii) static transfer switch, (iv) logic and control system, (v) battery bank, (vi) diagnostics and communications blocks. Depending on the capacity and the type of UPS, these blocks may have any thing from a few components to microcontroller or DSP based firmware.

In medium and large capacity online UPS systems a fair amount of modern power semiconductors such as monolithic Darlingtons, power MOSFETS, GTOs, and IGBTs are used currently, while the MCT devices hold a promise for the future. Most of these devices, mounted on specially designed heatsinks, etc., become the power control elements in the rectifier blocks and the inverter. Static switch is usually designed using state of the art thyristors etc., while diagnostic, communication, and control logic is based on simple logic blocks to microcontroller or DSP based systems.

All UPS systems have at least one large, low frequency, magnetic component, usually a transformer. Especially at high powers, even the latest PWM UPS systems require a number of traditional laminated iron magnetics. These are large, heavy, lossy, and expensive, and a great deal of research and development is directed at reducing the quantity and size of these magnetics. Many of the techniques described in this chapter require the use of high frequency magnetics.

The main advantages of higher frequencies are that acoustic noise can be reduced, and flicker components become smaller. Advances in the design of magnetic materials have not kept a pace with those of semiconductor devices and designs. It is the magnetic materials which are limiting the speed of high frequency developments, especially at powers greater than about 10kVA. In the next few sections individual blocks will be described.

7.3.1 Rectifiers and Battery Chargers

In low capacity (less than 20kVA) UPS systems, simple single or three phase rectifiers with no voltage control or minimal voltage control are used. However, these simple techniques are not usable with large capacity systems due to harmonic distortion and low power factor which is generally not acceptable by utilities.

The emphasis in all new rectifier designs is to achieve low current distortion at unity power factor. Single phase and smaller three phase UPS invariably use a diode bridge, followed by a high frequency switching stage. The control of this switching stage determines the input characteristics of the rectifier. Modern systems use active rectifiers with IGBTs as shown in Figure 7–8.

Such circuits promise active power (displacement) factor compensation, and the possibility of "intelligent" selection of input current characteristics, using IGBTs or Darlington transistors at output ratings from about 10kVA to 50kVA. Modern DSP based designs are sometimes used to achieve the mathematical processing demanded by these circuits, particularly in the high power systems.

Such rectifier circuits are intended for systems using a high voltage (700–800 volt) DC bus, and complement the transformerless inverter designs described later. However, batteries at these high voltages have inherent disadvantages, and batteries in the 400 to 500 volt range still predominate. These allow standard switchgear to be used, but require conversion up to the high voltages demanded by some inverter techniques. A typical three phase rectifier-battery charger system is illustrated in Figure 7–9.

One trend is towards the use of resonant switching in the high frequency stage, as shown in Figure 7–10. Such "soft switching," that is, switching at either zero voltage or zero current, reduces the problems caused by fast switching transients, which, in turn, reduces the RFI generated by these switching stages.

Most three phase UPS systems, especially above about 30 kVA, still use a 6 pulse SCR bridge as shown in Figure 7–11(a). This has the disadvantage of low power factor, and high distortion in the range of 30 percent. However filters are

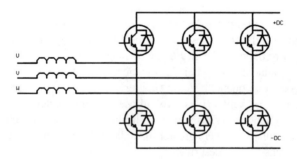

FIGURE 7–8 Three phase active rectifier with IGBT elements

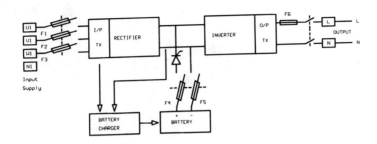

FIGURE 7-9 Typical three phase rectifier–battery charger

available to improve the power factor to above 0.9, and reduce distortion to about 10 percent.

In order to achieve lower distortion a popular technique is the 12 pulse rectifiers as per Figure 7–11(b). This solution, sometimes referred to as a double bridge, can result in distortion between about 12 percent to 5 percent depending on the use of a filter stage (McLennan 1994). Coupled with suitable input isolation transformers it could selectively reduce certain harmonic components and could provide complete galvanic isolation at the UPS input.

7.3.2 Inverters

Historically, inverter debate has focused on Pulse Width Modulation (PWM) and Quasi-Square-Wave (QSW) techniques. QSW is also known as DC Amplitude Control (DAC). PWM techniques are generally popular among low power inverter

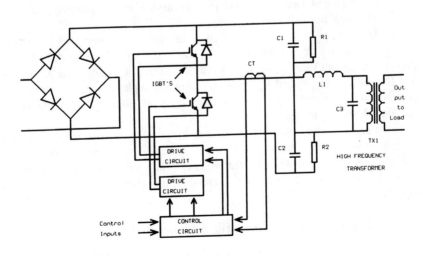

FIGURE 7-10 Resonant DC converter

214 POWER ELECTRONICS DESIGN HANDBOOK

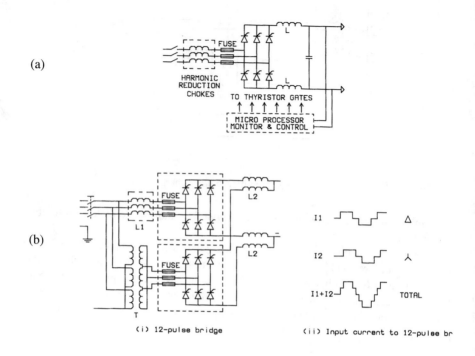

FIGURE 7-11 Charger configurations (a) Six pulse charger (b) Twelve pulse charger

designs, while DAC is the only static technique applicable at powers in excess of about 500 kVA (McLennan 1994). The discussion in the following sections is restricted to low power PWM techniques.

The basic switching action of inverter systems, irrespective of the power semiconductor switches used, is similar. The bridge usually consists of four main power devices per phase, two of which are switched on at any instant. See Figure 7–12(a). On closing S_1 and S_4 current will flow through the load. Conversely, closing S_2 and S_3, current again flows through the load. However, the polarity across the load is now reversed.

In practice, the switches are replaced by power semiconductors and the load by the primary of the output transformer. Figure 7–12(b) is a three phase inverter where transistors are used as power switches. The ease with which transistors can be switched on and off gives a considerable advantage over the thyristor based inverters.

7.3.2.1 Inverter Switching Principles

To explain the practical design of inverters, let's refer to Figure 7–13(a), where a dual IGBT module is connected to a DC voltage rail. When operating as a

(a)

(b)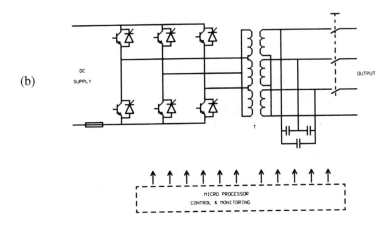

FIGURE 7-12 Inverters (a) Principle of inverter bridge (b) Three phase transistor Inverter

switch the circuit has two stable states of interest. The first is when Q1 is turned ON and Q2 is OFF, whereupon the inverter output is effectively connected to the +ve DC busbar; the second is when the situation reversed, where Q2 is ON and Q1 is OFF and the output is connected to the -ve busbar.

By alternating between these two states the output can be represented in the form of a square wave whose amplitude is dependent upon the DC busbar voltage and whose frequency is determined by the rate at which the transistors change from one state to the other. To ensure that the "off-going" transistor has sufficient time to turn off before the "on-coming" transistor is turned on there is a small delay between the removal of one drive signal and the application of the other; this period is known as the dead band (underlap or the dead time) and is in the order of one or two μS. Figure 7–13(b) depicts the inverter operation when the two transistors are driven by a true square-wave; that is, a gate drive signal having a 1:1 mark-to-space ratio. This produces a square wave output whose "mean" voltage, with respect to the -ve DC busbar, equals approximately 50 percent of the DC busbar voltage.

216 POWER ELECTRONICS DESIGN HANDBOOK

Similarly in Figures 7.13(c) and 7.13(d) the gate drive signals are shown producing output waveforms having mark-to-space ratios of 3:1 and 1:3 respectively, and the corresponding change in the average voltage is shown to be 75 percent and 25 percent of the DC busbar voltage, as expected.

These diagrams illustrate that the inverter "mean" output voltage can be varied anywhere between 0V and the full DC busbar voltage by controlling the mark-to-space ratio of the transistor drive signals.

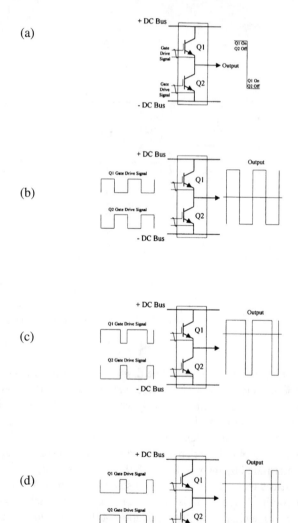

FIGURE 7–13 Illustration of PWM switching principles (a) Basic inverter switch (b) Inverter switching at 50% duty cycle (c) Inverter switching at 75% duty cycle (d) Inverter switching at 25% duty cycle

This principle can be extended to produce a sine wave output by driving the transistors with a drive signal as in Figure 7-14(a). However, it is necessary to filter the high frequency signal using a low pass filter between the inverter bridge output and the load as per Figure 7-14(b).

It is important to understand that the drive frequency remains constant and the output voltage is controlled entirely by altering the width of each pulse in the drive train. This technique is the most common inverter technique used in UPS inverters and is called Pulse Width Modulation (PWM). The 50 or 60 Hz AC output is extracted from the PWM waveform by the low pass filter which removes the high switching frequency. In practical systems the switching frequency could be between 2 kHz to over 20 kHz.

7.3.2.2 Selection of Transistors

In recent years manufacturers have been changing over to transistorized inverters compared to older generations which were based on SCRs. Conventional power transistors, Darlingtons, power MOSFETs, or IGBTs are used in modern inverters, depending on the capacity and the inverter design. The power handling capability of transistors is directly related to their physical size, as is their ability to switch at high speeds.

(a)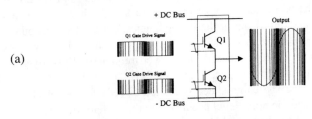

(b)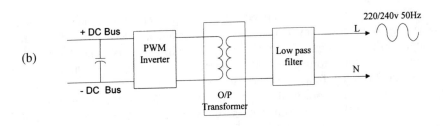

FIGURE 7-14 Pulse width modulation technique (a) PWM waveform to drive the transistors and corresponding output (b) Filter stage to remove high frequency components

In order to switch high load currents paralleling of transistors or connecting several inverter stages in parallel need be used. Both solutions, however, tend to reduce the reliability owing to the increased component count, and the difficulty in monitoring and controlling the parallel components.

For low power systems power MOSFETs and bipolar Darlingtons are used. When switching frequencies for the PWM waveform are selected between 2kHz to 5kHz monolithic Darlington modules could be used. With such devices noise levels of not more than 60 dBA could be achieved, which is lower than the noise level encountered in most computer rooms. In order to save costs many practical variations of the basic technique described in the previous sections are used in practical inverters.

IGBTs are the preferred transistors for modern UPS designs. They are significantly more efficient and are easier to control than any other power semiconductors. It is reported that IGBTs are commonly available for UPS applications up to 750 kVA without paralleling devices, while IGBTs suitable for UPS ratings up to 375 kVA are competitively priced and widely available from multiple sources (Hussain and Sears 1997).

Most UPS applications under 50kVA capacity use high frequency PWM techniques in their IGBT based inverters. These inverters could be switched at relatively high speeds such as 20 kHz. The high switching frequency helps improve output dynamic response and helps reduce the cost of magnetics and other components. While switching losses are significant at these frequencies, overall UPS efficiency in these power ranges are not the most important design goal. Larger UPS systems, over the capacity of about 50 kVA, generally use lower frequency PWM inverters. These switch at a significantly lower speeds between 2kHz to 5 kHz.

The importance of the switching speed is related directly to the selection of the output filter and the audible noise level produced by the system. If, for example, switching frequency could be above 15 kHz the frequency of the noise produced is above the audible range. The difficulty arises, however, when transistors need be selected, where high currents and a large DC rail voltage need be tackled at high frequencies. Given a nominal DC voltage of about 400 V the corresponding current for systems with ratings of 10–100kVA would be of the order of 30–300 A.

7.3.2.3 Practical Inverter Circuits

The full bridge circuit described could be implemented with many variations depending on the power levels involved and the available transistor modules. One example is shown in Figure 7–15 where the two IGBT pairs need not have identical high frequency performance. In this case module 1 is a high frequency type and the module 2 is a low frequency type. As this could save costs of the power semis it is frequently used in practical systems. In this case Q_1 and Q_2 are switched by a high frequency PWM waveform with a sine wave envelope as described earlier while Q_3 and Q_4 are switched using a 50 percent duty cycle line frequency waveform. PWM drive signal could be in the range of 10 to 20 kHz.

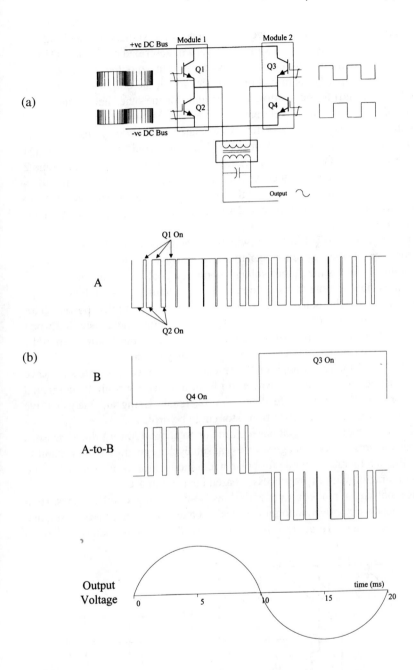

FIGURE 7-15 A practical inverter using IGBT pairs (a) Block diagram (b) Switching waveforms

220 POWER ELECTRONICS DESIGN HANDBOOK

The IGBTs are switched by two sets of gate drive signals. Q1 and Q2 are switched by the high frequency PWM drive waveform described earlier, and Q3 and Q4 are switched at the line frequency. Thus, considering module 2, for one half cycle of the UPS output Q3 is turned on and one side of the output transformer is referenced (connected) to the positive busbar; and, for the other half cycle, Q4 is turned on and the transformer is referenced to the negative busbar.

Figure 7–15(b) shows the waveforms associated and how they are "summed" at the output transformer primary. A dual IGBT package could be used for Q3/Q4 in the same way as for Q1/Q2. However, due to its lower switching rate, module 2 does not require such a high frequency specification as module 1. Therefore different types of devices could be used. Module 1 should have fast rise and fall times, to minimize switching losses, and module 2 should be a low $V_{ce(sat)}$ device to minimize conduction losses.

Referring to Figure 7–15(a), this switching arrangement together with the output transformer and filter capacitors generates a sine wave at the line frequency. By feedback control circuits the output voltage amplitude could be regulated.

It is important to note that due to reasons of matching between load and the inverter/filter stages, the deliverable load power is within a limited power factor range, usually between 0.65 to 0.85. When loads are beyond these specified worst case power factors voltage regulation as well as total harmonic distortion may be affected or tripping or changeover to static by pass may occur.

By the use of microprocessor or DSP blocks a low-power accurate sine wave could be generated and the output waveform from the inverter could be compared in order to produce the necessary feedback control signals. Using suitable gate drive circuits gates or the bases of power transistor could be driven.

Pulse width modulation systems are used up to about 400 kVA due to reasons of low acoustic noise, good voltage waveform, sub cycle regulation and availability of many silicon integrated circuits for pulse width modulation control, which could work in conjunction with microprocessor based control blocks.

The emphasis in modern inverter designs is towards higher efficiencies. High efficiencies lead to improved autonomy times on battery, reduced power losses, and lower power rectifiers. To achieve high efficiency, the optimum balance between

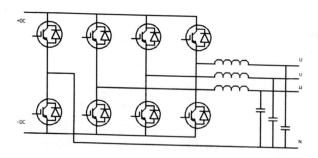

FIGURE 7-16   Transformerless 4 pole inverter bridge

switching losses (which increase with increasing switching frequency) and conduction losses (which decrease with increasing switching frequency), must be chosen. Modern designs are using IGBTs particularly since the new generations of IGBT have progressively lower losses and the ability to switch higher DC voltages. IGBTs are under intense competitive development, and it is unlikely that any new UPS designs will be released using power Darlington transistors.

Recent research papers have proposed the use of transformerless inverters using a 4 pole bridge, as illustrated in Figure 7–16. Provided losses in the switching devices can be minimized, and other practical design difficulties overcome, eliminating the transformer will further increase efficiency, and reduce the cost and size of the UPS. For use in 415 volt, three phase systems, transformerless designs require an inverter DC bus voltage of 700–800 volts. This complements the active rectifiers described earlier.

Despite some recent trends towards lower PWM frequencies (McLennan 1994), the general trend is still towards higher and higher frequencies, encouraged by associated advances in large magnetic cores. The cost of silicon is reducing rapidly, the cost of copper and iron are increasing, which may see more complex inverter designs specifically to reduce or eliminate transformers and chokes.

7.3.3 Use of Microcontrollers and Output Voltage Control

In most modern UPS systems a microcontroller is used as the central control element for many primary functions such as: (i) generation of PWM drive waveform, (ii) monitoring the O/P voltage, (iii) inverter output current for overload and short circuit protection, (iv) phase and frequency locking of output and the input power supply, (v) supplying the gate driver stages with properly timed waveforms, etc., in addition to charger control functions. Also, it could carry out many secondary functions such as: (i) providing information display, (ii) fault detection, (iii) battery health monitoring, and, (iv) miscellaneous "intelligence" functions.

Figure 7–17 depicts a block diagram of inverter with a simplified block diagram of a control stage. The variable PWM drive signals and line frequency IGBT drive signals are generated digitally by the processor system on the microcontroller block. The precise PWM pattern used is determined by an error detection circuit which monitors the output voltage and either increases or decreases the mark to space ratio of the PWM drive signal, as necessary in order to maintain the correct output voltage.

The processor system monitors the frequency and the phase of the utility voltage for maintaining synchronism between the inverter output and the bypass supply (in-coming mains). Also the control system monitors the inverter current for providing overload and short circuit protection.

Synchronization is carried out by monitoring the zero cross-over point of the utility voltage, which provides the processor with both frequency and phase information. The utility frequency is measured to determine whether or not it is within the permissible synchronizing window.

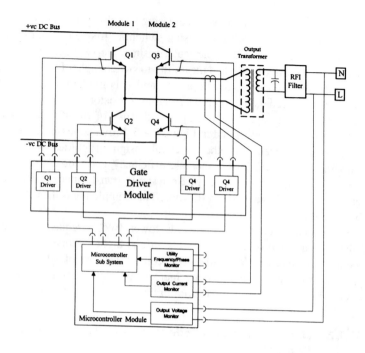

FIGURE 7-17 Inverter stage with processor control

If it is within the window then the line frequency inverter drive signal is synchronized to the zero crossover signal which keeps the inverter output synchronised with the bypass supply. If the utility frequency is outside the synchronizing window the processor reverts to its internal clock system to provide the line frequency drive waveforms. The current sense signal could be used for various overload protection functions, both software and hardware controlled, depending on the severity of the sensed error. Output capacitors, and RFI filer, etc., are used to provide a clean, harmonic, and interference free sine wave output, which is continuously monitored by the microcontroller subsystem.

7.3.3.1 Gate Drivers

The gate drivers are used to provide the necessary gate drive signals, when the gate needs a voltage signal (or a charge release) with respect to its corresponding source terminal (of a MOSFET) or the emitter terminal (of an IGBT). In general, the necessary signal may be with or without reference to the -ve voltage rail and creates the need for complex circuits. For this reason, most UPS inverter systems use special gate driver modules or gate driver boards.

As depicted in Figure 7–17 a gate driver block contains an individual drive circuit for each IGBT. This module is responsible for providing the necessary drive power and galvanic isolation between the signals leaving the microcontroller block and the IGBT gate/emitter terminals.

7.3.4 Static Switch

The inclusion of a static switch is an absolute necessity for systems in the medium power (10 to 100KVA) range. It comprise a pair of "back to back" thyristors in each phase of the bypass supply (Figure 7–18) with a parallel contactor.

In the event of overload or failure of the inverter the load would be transferred from the UPS to the mains supply by means of the static switch. In order to achieve this, certain parameters must be met. The two sources, inverter and bypass, must be synchronized and the voltage must be within reasonable limits (± 10 percent). Under normal operating conditions the inverter will synchronize with the bypass supply, provided the bypass is within acceptable limits to the load. The tolerance is normally ± 1 percent of nominal frequency (the tolerance range is however, selectable over a range of ± 0.5 to ± 2 percent). To ensure no break transfer the phase must be within $3°$.

If the transfer has been initiated by an overload condition, on the cessation of the overload the load would again be transferred to the inverter. In this manner, the static switch would supply the inrush current normally associated with initial switch on, again avoiding the necessity to oversize the unit.

Theoretically, on-line UPSs provide near perfect isolation between the line input and the load for all types of line disturbance. However, this isolation may be compromised because most on-line UPSs incorporate a static bypass switch, which under certain circumstances connects the load directly to the line input, thus bypassing the unit's rectifier and inverter.

Bypassing the rectifier and inverter is necessary to cope with high transient load currents, such as the inrush current drawn by equipment each time you switch it on. On its own, the inverter is usually unable to cope with these transient currents.

The static bypass switch is also activated if the inverter itself fails for any reason. While this switch is activated, the load is unprotected from any line disturbances unless the bypass circuit itself includes line-conditioning circuitry. In addition, most static bypass switches have a transfer time of a few milliseconds so that each time the switch is activated, load may be deprived of part of an input cycle.

7.4 UPS Diagnostics, Intelligence, and Communications

As in the case of many general industrial electronic systems, modern UPS systems are designed with simple remote alarm panels with associated communication interfaces to sophisticated communication interfaces for network hubs and file

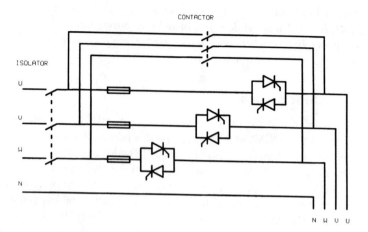

FIGURE 7-18 Arrangement of static switch

server locations, etc. Simple contact signals indicate "on battery" or "low battery" at remote alarm panels at significantly low additional costs.

For communication over short distances (up to about 100m), the remote facility could be hardwired. Over long distances, information could be relayed via modems and phone lines. Most manufacturers are now utilizing the diagnostic and communication facilities to provide remote detailed diagnostics from service locations or the network manager locations, etc.

7.4.1 Intelligent UPS Systems

An intelligent UPS system has the capability to communicate its status back to the network. With network management software and an intelligent UPS, the network manager can remotely inquire about the UPS operation and the current power condition at the computer site. In addition, the manager can remote control the intelligent UPS and the file server's specific response to a power problem.

Until relatively recently, UPS systems had no "intelligence," meaning that any remote monitoring or control functions could only be obtained by special proprietary circuitry provided by the UPS system manufacturer. The development of client/server network technology introduced the need for communications far beyond what was envisioned previously even for large multi-module three phase systems.

The monitoring and control of very large complex networks developed the need for an open systems standard interface software package that can work with UPS systems of different manufacturers across a variety of diverse operating systems, and in addition, provides system security.

With the simple network management protocol (SNMP) getting widely accepted, UPS designers have developed the communication interferes to interact with SNMP suitably. SNMP includes the capability to monitor and/or control from a central console the network elements (such as servers, gateways, bridges and routers, etc.) as well as a UPS system.

7.4.2 Levels of UPS Intelligence

There could be several levels of intelligence (Katzaro 1993), as follows:

7.4.2.1 Level 1

The simplest communications package, UPS monitoring, may be used in a local departmental network. In this case the UPS is connected to the file server or host, allowing the file server to be shut down in the event of a power disturbance. The basic non-intelligent UPS has simple contact closures that send two messages to the file server or host indicating an "on-battery" or "low battery" condition. Based on the messages, the network can broadcast the status to the users or initiate an operating system shutdown procedure.

7.4.2.2 Level 2

Where there are several departmental LANs under the control of one manager, this manager can use the software from any node on the network to monitor and control any UPS system and to switch any device to reset and reboot. By using the RS-232 protocol, true two-way communication is possible and a much valued feature.

7.4.2.3 Level 3

In an enterprise network where a number of LANs have been joined together, all critical elements on the network should and can be centrally managed through SNMP.

7.4.3 SNMP Products

An example of such SNMP products is American Power Conversion's PowerNet SNMP Manager. It provides:

(a) A graphical user interface that monitors all UPS units on the network
(b) Notification of power alerts to network managers
(c) UPS diagnostics
(d) Automatic data-save and server-shutdown features upon power failure
(e) Scheduled automatic UPS testing
(f) Rebooting of locked-up network devices

The American Power Conversion package supports Unix with HP Open-View and IBM NetView/6000. Similar network-management UPS software packages are available from Best Power, Clary, and Deltec, with support for Netware, Unix, DOS/Windows, and OS/2 (Travis 1995).

7.5 UPS Reliability, Technology Changes and the Future

Over the past 10 years, the UPS industry has made substantial improvements in the reliability of its products and systems. The most significant factors effecting the reliability increase are a reduction in component count, higher quality component availability, and the industry's movement toward "proofing" new designs before production. Figure 7–19 shows a generalized curve indicating the basic component and topology changes over the years that have led to significant reliability enhancements, primarily due to the component count reduction (Burgess 1991).

At all but the highest power levels, the use of pulse width modulation in place of step wave modulation has permitted the number of power switches to decrease by up to a factor of four.

The evolution of high power Darlington modules and recently the IGBT modules has given the power electronics circuit designer the ability to explore and implement many alternatives for power switches. In addition, these devices have permitted operating frequencies to increase by two orders of magnitude at moderate power levels (60kVA) and up to three orders of magnitude for the desk top products.

Users can expect future UPS offerings to follow the trend of other electronic commodities; smaller, faster, and cheaper. As integration increases, the tendency toward smallness will only increase. Pushing that trend is the fact that several UPS units available now use µPs and ASICs in their control circuitry.

Both UPS hardware and software promise to undergo major changes in the future. For example, all UPSs now use lead-acid batteries. The main reasons for the choice are cost and power density.

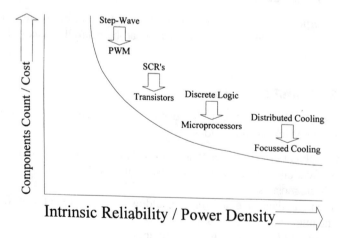

FIGURE 7–19 Technology impact on UPS systems (Source: Burgess 1991)

When exotic battery technologies, such as zinc-air and lithium-based designs, decrease in cost, it is likely they will gradually displace lead-acid units. Other hardware-based improvements expected are more efficient battery chargers and inverters. Power-switching devices are steadily improving in on-resistance, and those improvements will result in lower power losses. As power losses decrease, the use of on-line UPS units will probably increase because of their allure of zero transfer time and simple redundancy.

References

1. Katzaro, P. "Using Intelligent UPS Systems in the Distributed Processing Environment." Power Quality Conference Proceedings, Oct 1993, pp 77–92.
2. Folts, D. C. "The Response of Advanced Single Conversion UPS to Power Disturbances." Power Quality Conference Proceedings, Oct 1993, pp 93–104.
3. Gupta, S. and H. Handler. "Novel Phase and Frequency Locked Loops Scheme for Utility Interactive UPS." Power Conversion Conf. Proceedings, Oct 1990, pp 45–53.
4. Rando, R. "AC Triport—A New Uninterruptible AC Power Supply." Proccedings of the International Telephone Energy Conference, 1978, pp 50–58
5. Handler, H. and V. Rangaswamy. "Characteristics and Performance Data of Triports." Battery Conf. Proceedings, 1989,
6. McLennan, A. A. "Static UPS Technologies." IEE Colloquium, Feb 1994.
7. Burgess, W. "Evaluation of Reliability in UPS Products." Power Quality Proceedings, Sept. 1991, pp 198–206.
8. Travis, B. "Paranoid about Data Loss? Choose Your UPS Carefully." *EDN*, Sept 14, 1995, pp 131–138.
9. Hussain, A. J. and J. Sears. "UPS Semiconductor Technologies." *Power Quality Assurance*, March/April 1997, pp 12–18.

Bibliography

1. Stewart, W. N. "Selecting a UPS for Optimum Results." *PCIM*, December 1988, pp 39–41.
2. Peter, H. "Units Power PCs in Rough Times." *EDN*, May 25, 1989, pp 73–84.
3. Meirick, R. P. "The Evolution of Uninterruptible Power Supplies." *PCIM*, Oct 1989, pp 52–54.
4. Patrick, H. "Battery Backed Switcher Provides UPS Capability." *PCIM*, Oct 1989, pp 56–59.
5. Stewart, N. "Power Problems: Selecting a UPS." Electronics Test, July 1990, pp 26–29.
6. Stephens, N. O. "UPS Rating System Formalizes Purchasing Decisions." *Power Quality* July/August 1991, pp 5–52.
7. Sadayoshi, Seiji, et al. "Power Transistorised Three Phase Uninterruptible Power Supply." Fuji *Electric Review*, Vol 29, 1983, pp 59–65.
8. Sam, D. "Low Cost UPS Guards Against Data Loss." PCIM, July 1992, pp 32–35.
9. David, J. "Industrial and Process Control Systems Have Unique UPS Requirements." *Power Quality*, April-June 1992, pp 25–32.
10. Byrne, J. A. "Powering Tomorrow's Telecomputer Networks: A Challenge to Power System Designers." *PCIM*, Aug 1994, pp 86–96.
11. Platts, J. and J. St. Aubyn. *Uninterruptible Power Supplies*, Peter Perigrinus Ltd., 1992.
12. Bowler, P. "UPS Specifications and Performance." IEE Colloquium, Feb 1994, pp 1–13.
13. Merlin Gerin Inc. "High Power UPS Systems Design Guide." Document No. CG0035E/1.

14. Clemmensen, M. and C. W. Seitz. "Current State of UPS Technology." IEEE Transactions, IAS Conference, 1986, pp 1029–1035.
15. Wells, B. M. "Magnetic Components in UPS." IEE UPS Colloquium, Feb 1994.
16. Taylor, Brian E. "The Wrong and Wrong Reasons for UPS Requirements." Proceedings of Power Quality/ASD Conference Proceedings, Oct 1990, pp 55–63.
17. Raddi, W. J. and R. W. Johnson. "A Utility Interactive PWM Sine Wave Inverter Configured as a High Efficiency UPS." International Telecommunications Conference, IEEE Transactions, 1982, pp 42–45.
18. Larkins, G. P. "Selection, Application and Installation Considerations for Static Uninterruptible Power Supply Systems." Power Quality Conf. Proceedings, October 1993, pp 105–119.

CHAPTER **8**

Energy Saving Lamps and Electronic Ballasts

Co-Authors: Anil Gunawardana and Nalin Wickramarachchi

8.1 Introduction

Ever since the first energy crisis that the world faced during 1970s (due to a sudden and unexpected rise in the price of petroleum fuel), the electricity industry has been trying to meet the world's burgeoning energy needs by building more power plants that do not depend on oil or by looking for other non-conventional energy sources such as solar power. During the 1990s, however, a new concept called "negawatts"—the idea that the investment in energy conservation will often yield higher returns than investment in new power stations—is gaining popularity. According to this view the electricity demand could be limited by matching an appropriate and efficient technology to each energy utilization task.

Electric lamps are a case in point. A century after its invention the electric filament lamp is still one of the world's most popular ways of providing artificial illumination in industry as well as in households, despite the fact that the filament lamp yields comparatively the least light output for a given amount of electrical energy input. This figure known as the luminous efficiency or efficacy, hardly has been improved by any new technology as far as the filament lamp is concerned.

The other most popular source of electric lighting is the fluorescent lamp which uses the principle of an electric arc discharge through a gas at a low pressure to produce visible light. The illumination by the principle of gas discharge has been in existence for more than fifty years and almost all new research and development in the area of improving efficacy of lighting has been concentrated broadly on the fluorescent lamp technology. For instance, it has been estimated in the United States that the fluorescent lamps produced by a factory which costs eight million dollars to build will save electricity worth of one billion dollars, equivalent to the cost of a 700MW power plant.

Thus the phrase "energy saving lamps" is basically synonymous with the new technology being developed for the improvements in fluorescent lamp technology. Particularly, the development of low-wattage fluorescent lamps together with highly efficient electronic ballasts (the auxiliary circuit required to control the operation of a gas discharge lamp) is the main focus of the lighting industry today.

This chapter provides an overview of these new energy saving techniques as applied to fluorescent lamps. The use of modern Application Specific Integrated Circuits (ASICs) in practical electronic ballasts, as well as some magnetic ballast technologies are discussed. This chapter also provides a set of definitions, units, and measures for the purpose of evaluating and comparing the performance of different types of lamps.

8.2 Gas Discharge Lamps and High Intensity Discharge Lamps

8.2.1 The Fluorescent Lamp

The fluorescent lamp which was first developed in 1930s consists of a tube that is coated on the inside with a fluorescent powder, or phosphor. The tube contains mercury vapor at low pressure with a small amount of an inert gas to assist ignition of discharge. Two electrodes are placed at either end of the tube and are designed in such a way so as to operate as either "hot" or "cold" cathode lamps.

Hot cathode lamps contain electrodes made of coated tungsten filaments and are usually heated to an electron emitting temperature before the arc strikes. The heated cathodes facilitate a low voltage drop of about 10 to 12 volts at the electrodes, saving approximately 3W per lamp.

Cold cathode lamps use coated electrodes of iron or nickel. The voltage drop at the electrodes of these lamps is relatively high (50V and above) but they exhibit a longer life due to low operating temperatures.

The operation of a fluorescent lamp consists first of establishing a sustained electric arc between the two cathodes. The impact of these electrons with the atoms of mercury vapor produces mostly invisible ultraviolet light which is then converted into visible light by the phenomenon of fluorescence of the phosphor coating on the tube. The chemical composition of the phosphor coating is, therefore, mostly responsible for the color of the light produced and also partly for the efficacy of the lamp.

The standard fluorescent lamp with a conventional halophophor coating produces a whiter color than the incandescent lamp. Adding a thin coat of more expensive tri-phosphor can improve the color rendition and increase the efficacy.

In general, the fluorescent lamp is a widely used light source with a good efficacy of about 90 lux/watt without considering the power losses in the ballast. When these losses are included the efficacy drops to about 75 lux/watt, a figure which is still far better than that of an incandescent lamp (see Figure 8–1).

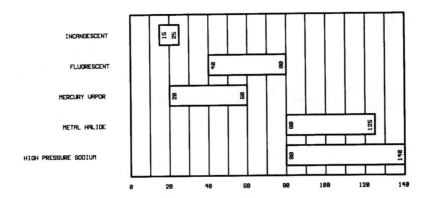

FIGURE 8-1 Comparison of lamps, in terms of lumens per watt

8.2.2 Compact Fluorescent Lamps (CFL)

The compact fluorescent lamp does not differ in its operating principle from the standard fluorescent lamp, yet the CFL has been designed to address some of the fundamental objections to the widespread application of the linear fluorescent tube in many residential, commercial, and industrial applications. The bulky magnetic ballast, flickering of light and the sometimes audible noise generated by the magnetic ballast were some of the reasons for the unpopularity of the fluorescent tube as a general purpose light source.

The CFL overcomes flickering by operating the lamp at the kHz frequency range and gets rid of the need for an external ballast by incorporating an all-electronic ballast into the base of the fluorescent tube. Thus, CFLs are intended and capable of directly replacing the incandescent lamps without any external auxiliary devices.

Figure 8–2 shows the basic block diagram of a compact fluorescent lamp. Note that the electromagnetic interference (EMI) filter and power factor control blocks are necessitated by the presence of electronics for AC/DC conversion DC/AC high frequency conversion circuits inside the package.

8.2.3 High Intensity Discharge (HID) Lamps

This is a general term for a group of lamps including mercury vapor lamps, metal halide lamps, and high pressure Sodium lamps.

The Mercury Vapor Lamp is a high pressure electric discharge lamp in which the major portion of radiation is produced by excitation of mercury atoms. Switching on the normal mains voltage is not sufficient to start the discharge between the main electrodes. It can, however, start over the very short distance between the main and auxiliary electrodes, the auxiliary electrode being connected to the lamp terminal through a high resistor for limiting the current.

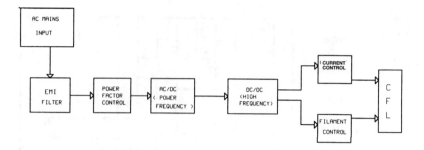

FIGURE 8-2 Block diagram of a CFL

The initial discharge is in the small amount of argon present. The discharge now spreads rapidly until it takes place between the main electrodes. The argon discharge warms up the tube and vaporizes the mercury. The discharge then takes place in the mercury vapor and the argon has negligible effect. The efficacy of the lamp is about 60 lux/watt.

The Metal Halide Lamp is an electric discharge lamp in which the light is produced by the radiation from an excited mixture of metallic vapors (mercury and products of the disassociation of halides). Their construction is similar to high pressure mercury lamps, a number of iodides are added to fill the gaps in the light spectrum, improving the color characteristic of the light. Their efficacy is also higher (up to 80 lux/watt).

The sodium Lamp contains neon in addition to the metallic sodium at low pressure. Heat is produced by an initial neon discharge. This causes the sodium to discharge giving a sodium–yellow color. The color is caused by the excitation of sodium vapor. It takes about ten minutes for full light to be reached. A development of this is the high pressure sodium lamp which at high pressure has a widened spectrum to give an adequate cover of all colors. sodium vapor lamps have a very high efficacy of up to 150 lux/watt.

Fluorescent Lamps are popular because they provide longer lifetimes than incandescent, and consume less energy. Further, their low intensity even illumination is preferred in almost all indoor conditions. The high intensity discharge lamps are used primarily in outdoor conditions to light large areas such as streets, parking lots, etc.

8.3 Introduction to Ballasts

Fluorescent lamp ballasts are devices installed in fluorescent lamp fixtures in order to regulate the voltage and current provided to the lamps. The functions of a ballast in a fluorescent lamp circuit are two fold. First, it must provide a suitable striking (ignition) voltage across the bulb at starting such that an electric arc can sustain between the electrodes afterwards. Secondly, the ballast is responsible for limiting the current flow across the lamp during the normal operation of it. These two requirements of a ballast can be explained from a typical impedance-time characteristics curve of a gas discharge lamp shown in Figure 8–3.

As the initial impedance is high, the striking voltage required to ignite the arc would also be higher than the normal operating voltage of a fluorescent lamp. Immediately after the lamp is struck, the impedance drops to its minimum value, representing a negative resistance characteristics that need some form of current limiting to prevent lamp destruction from excessive current.

While the early magnetic (inductor type) ballasts fulfilled the two necessary requirements of a ballast, the modern electronic ballasts can perform many other functions such as resonant operation, protection from lamp shutdown, failure or removal, and dimming operation, etc. These and other such techniques will be discussed in detail in forthcoming sections.

It should also be noted that while a fluorescent lamp by itself is purely a resistive load, the incorporation of a ballast (either magnetic or electronic type) with it can give rise to potentially objectionable conditions such as low power factor, high order harmonics, and electromagnetic interference. As we will see later in this chapter, many IC manufacturers have come up with advanced products that solve these problems quite satisfactorily.

8.4 Some Definitions and Evaluation of Performance

The primary measure of effectiveness of an electric lamp is its total luminous flux output for a watt of input power. For the purpose of comparison of performance between different light sources, firm definitions of the terms involved are necessary.

8.4.1 Luminous Flux

The total quantity of visually evaluated radiation (i.e., light) emitted per second by a light source is termed luminous flux and is measured in lumens. The term

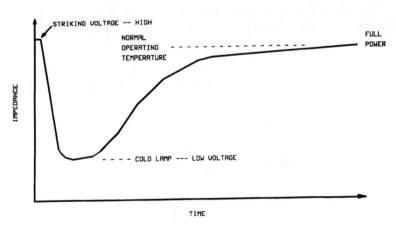

FIGURE 8–3 Impedance-time characteristics of a discharge lamp

"visually evaluated radiation" refers to the fact that humans are capable of seeing only a part of the spectrum of electromagnetic radiation.

Moreover, the sensitivity of the human eye differs widely at different wavelengths within this band of frequencies. The luminous flux as measured in lumens takes both these factors into account and, thus, there is no direct correspondence between radiation energy emitted per second of a light source and its lumens output.

8.4.2 Luminous Efficacy

The luminous flux output of an electric lamp for a watt of input power is defined as the luminous efficacy of a lamp. It is usually expressed in lumen/watt:

$$\text{Luminous Efficacy} = \frac{\text{Luminous Flux}}{\text{Power Input}}$$

Luminous efficacy is sometimes also referred to as lumens-per-watt or lpw rating of a lamp. According to current standards, the luminous efficacy of a fluorescent lamp must be measured inclusive of the power consumption of the ballast.

8.4.3 Current Crest Factor

The Current Crest Factor is the ratio of the peak lamp current to the rms current.

$$\text{Current Crest Factor} = \frac{\text{Peak Current}}{\text{RMS Current}}$$

It is a consideration of the lamp current wave form. The maximum crest factor recommended by lamp manufacturers so as not to degrade lamp life is around 1.7.

8.4.4 Ballast Factor

The Ballast Factor is the ratio of the light output of the lamp with the ballast under test to the lamp light output with the ANSI (American National Standards Institution) reference ballast.

$$\text{Ballast Factor} = \frac{\text{Lamp light output with test ballast}}{\text{Lamp light output with reference ballast}}$$

8.4.5 Ballast Efficacy Factor (BEF)

The BEF is the ratio of the ballast factor to the input power of a lamp-ballast system. The BEF is application specific and cannot be used to compare different applications.

$$\text{Ballast Efficacy Factor} = \frac{\text{Ballast Factor}}{\text{Input Power}}$$

8.4.6 Total Harmonic Distortion (THD)

The THD measures the quality of the current waveform produced by a ballast. The current drawn by a ballast in most cases has a nonsinusoidal waveform and thus can be considered as a series of high order harmonics (i.e., waveforms with frequencies that are multiples of input line frequency) superimposed on the fundamental current waveform. The extent of the presence of such harmonics are measured by THD as defined below.

$$\text{THD} = \frac{(h_2^2 + h_3^2 + h_4^2 + \ldots)^{\frac{1}{2}}}{h_1} \tag{8.1}$$

where each term h_i refers to the rms value of the ith harmonic in the current waveform, and h_1 refers to the rms value of the fundamental component.

8.5 Conventional Ballasts

In the conventional ballast circuit shown in Figure 8–4, the high voltage kick needed to strike the lamp is obtained form the inductor and a bi-metallic switch that also supplies filament current when the contacts are closed. The heated filaments emit space charges that lower the ionization voltage of the mercury vapor within the lamp for easy starting (Mortimer 1994). As the length of the arc tube increases, ionization voltages also increase, requiring ballast to provide stepped up operations voltages as well as higher striking voltages. As a consequence, conventional two and four feet fluorescent lamp ballasts use bulky step up high reactance transformers with output windings to drive two or more lamps.

This magnetic type of ballast is based on a coil of wire surrounding an iron core. Such magnetic ballasts of traditional design are also known as "core and coil ballasts." While a standard magnetic ballast dissipates about 20 percent of the total power, a more efficient magnetic ballast will limit this loss to 12 percent or less. The magnetic ballast is responsible for some harmonic generation due to the non-linear magnetization characteristic of iron.

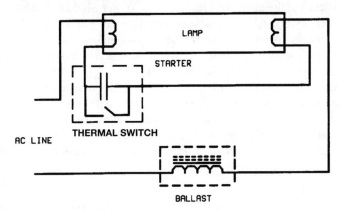

FIGURE 8–4 Basic arrangement of a conventional ballast

The inductance of the magnetic ballast presents poor power factor, typically around 0.5, which needs to be compensated for. Power factor compensation can be done by the use of a capacitor. Even after compensation, low quality magnetic ballasts will have a power factor of around 0.9 due to the relatively high THD of 20–30 percent. The conventional line frequency magnetic ballasts are associated with the following drawbacks.

(i) Flicker from 50/60 Hz power mains,
(ii) Significant size and weight,
(iii) Low power factor, non-sinusoidal current wave forms, and
(iv) Difficulty for dimming.

8.6 High Frequency Resonant Ballasts

Electronic high frequency resonant ballasts are being increasingly used to drive fluorescent lamps due to their improved energy efficiency, longer lamp life, dimming capabilities, lighter weight, and ability to eliminate flicker.

One of the earliest examples of the electronic operation of fluorescent lamps is found in a 1954 design manufactured by Delco for use in buses. This early electronic ballast was designed to operate six lamps at a total output power of around 140 watts. It operated at around 3000 hertz and was quite large (of the order of 1500 cubic inches). Improvements in semiconductor devices allowed the first practical high frequency ballast to be produced by Triad-Utrad in 1967. These ballasts were simple current fed self oscillating inverters and were also designed for DC input applications (Mortimer 1994).

Electronic ballast circuits have recently undergone a revolution in sophistication from the early bipolar designs of ten years ago. Partly, this has been brought about by the advent of power MOSFET switches with their inherent advantages in efficiency. Most electronic ballasts use two power switches in totem pole (half-bridge) topology and the tube circuits consist of L-C series resonant circuits with the lamp(s) across one of the reactors. Figure 8–5 shows the basic topology.

The switches in the circuit of Figure 8–5 are power MOSFETs driven to conduct alternatively by two secondary windings on a current transformer. The primary of this transformer is driven by the current in the lamp circuit operating at the resonant frequency of L and C. The circuit is not self starting and must be pulsed started by a Diac connected to the gate of the lower MOSFET.

After the lower switch is turned on, oscillation is sustained and a high frequency square wave (30–80 KHz) excites the L-C resonant current. The sinusoidal voltage across C is magnified by the quality factor (Q) at resonance and develops sufficient amplitude to strike the lamp, which then provides flicker-free illumination.

This circuit has been the standard electronic ballast for many years despite the following inherent shortcomings:

(i) Not self starting,
(ii) Poor switching time leading to higher power losses,

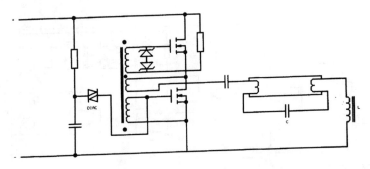

FIGURE 8-5   Electronic ballast using transformer drive

(iii) Labor intensive to manufacture (due to torroidal current transformer, etc.)
(iv) Not amenable to dimming, and
(v) Expensive to manufacture.

8.7   The Next Generation of Ballasts

The limitations of the basic electronic ballast circuit design and the need for more efficient lighting systems, coupled with the availability of power MOSFET switches, have created a push for small, efficient, low weight driver ICs. For example International Rectifier's IR2155 self oscillating power MOSFET/insulated gate bipolar transistor (IGBT) gate driver, is one of the first in a family of power ICs, tailored to electronic ballasts for fluorescent lighting, partly because of its small size and low cost (around $2 per 80 units and $1 per 50,000 units).

These power ICs can drive low and high side MOSFETs or IGBTs from logic level ground referenced inputs. They provide offset voltage capabilities up to 600 volts DC and, unlike driver transformers, can provide super-clean wave forms of any duty cycle (0-99%). A functional block diagram of the IR 2155 is indicated in Figure 8–6. These drivers have two alternative outputs, so that a half bridge or totem-pole configuration of MOSFETs can produce a square wave output. A very useful feature of self oscillating drives is their capability to synchronize the oscillator to the natural resonance of an L-C fluorescent lamp circuit. Figure 8–7 shows the concept of an electronic ballast using an IR2155 driver.

The IR2155 provides the designer with self oscillating or synchronized oscillating function, merely with the addition of R_T and C_T components. The IR2155 MOS gate driver also has internal circuitry which provides a nominal 1 microsecond dead time between outputs and alternating high side and low side outputs for driving half bridge power switches. When used in self oscillating mode, the frequency of oscillation is given by:

$$F_{OSC} = \frac{1}{1 \cdot 4 R_T C_T} \qquad (8.2)$$

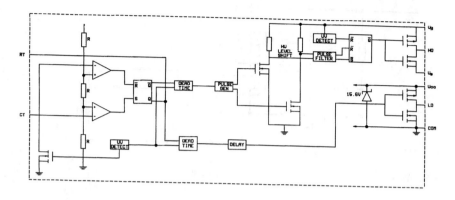

FIGURE 8-6 Functional block diagram of IR 2155 (Reproduced by permission of International Rectifier, USA)

Note the synchronizing capabilities of the IR2155 driver. The two back to back diodes in series with the lamp circuit are effectively a zero crossing detector for the lamp current. Before the lamp strikes, the resonant circuit consists of L, C1, and C2 all in series. C2 has a lower value than C1 so it operates at a higher AC voltage than C2, and in fact, it is this voltage that strikes the lamp.

After the lamp strikes, C2 is effectively shorted by the lamp voltage drop, and the frequency of the resonant circuit is governed by L and C1. This causes a shift to a lower resonant frequency during normal operation, again synchronized by sensing the zero crossing of the AC current and using the resonant voltage to control the IR2155 oscillator. A practical ballast circuit using the IR2155 integrated circuit, which is capable of driving two 4 foot tubes, is indicated in Figure 8-8.

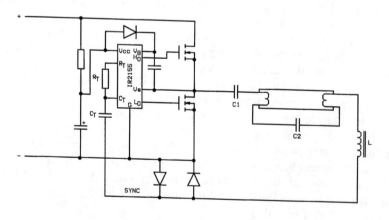

FIGURE 8-7 Electronic ballast using IR2155 driver (Courtesy of International Rectifier, USA)

Energy Saving Lamps and Electronic Ballasts **239**

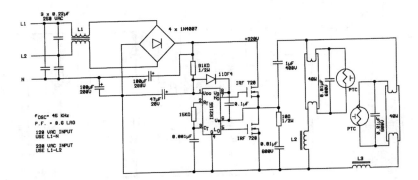

FIGURE 8-8 "Double 40" ballast using the IR 2155 oscillator/driver (Reproduced by permission of International Rectifier, USA)

One of the drawbacks of this circuit is the low power factor and high harmonic current. The circuit in Figure 8–7 accepts 115 volts or 230 volts AC 50/60 Hz input to produce a nominal DC bus voltage of 320 volts DC. Since the input of the rectifiers conduct only near the peaks of the AC input voltage, the input power factor is approximately 0.6 lagging with a non-sinusoidal current wave form.

8.8 Power Factor Correction and Dimming Ballasts

For electronic ballasts, it is possible to provide power factors exceeding 0.95 by using a boost topology operating at a fixed 50 percent duty cycle. Using the IR2155 driver, it is also possible to provide dimming merely by changing the duty cycle and hence the boost rates (Wood (April) 1994), as illustrated in Figures 8–9 and 8–10, respectively. Power factor correction is discussed in more detail in the next chapter.

8.9 Comparison of Compact Fluorescent Lamps Using Magnetic and Electronic Ballasts

The electronic ballast has many advantages over magnetic ballast. These include flicker elimination, low noise, longer ballast life, and of course, energy savings. The energy saving potential of electronic ballasts more than makes up for the additional cost they initially incur. These energy savings can be seen through lower power consumption, and indirectly, in the temperature of the ballast itself.

Electronic ballasts are not without their problems. The total harmonic current distortion is a real concern for engineers. Electronic ballasts can have a THD which far exceeds those of magnetic type ballasts. High harmonic levels have been linked to problems including capacitor bank failures, overheating of transformer windings,

240 POWER ELECTRONICS DESIGN HANDBOOK

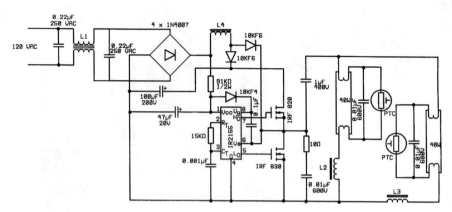

FIGURE 8-9 Ballast with active power factor correction

excessive neutral currents, de-rating of transformers, and misoperation of utility protective relaying. They have also been known to disturb sensitive electronic equipment which requires a clean sinusoidal wave form (Datta 1994). The results of a comparative analysis carried out on a number of samples of compact fluorescent lamps (CFL) both with magnetic and electronic ballasts as well as with integral and separate ballasts are shown in Figure 8-11.

A study (Lucas and Wijekoon 1995) has shown that most available low cost CFLs could have a very poor power factor. In particular, it has been shown that CFLs with magnetic ballasts can have power factors as low as 0.4 lagging on account of a highly inductive ballast, but they do not contribute to a high degree of harmonics. On the other hand, CFLs with electronic control gear have an almost equally poor effective power factor, mainly on account of harmonics caused by their power electronics.

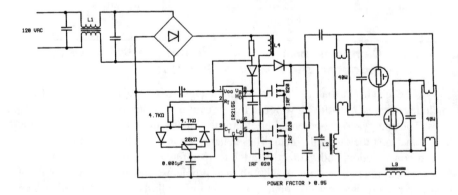

FIGURE 8-10 Dimming ballast

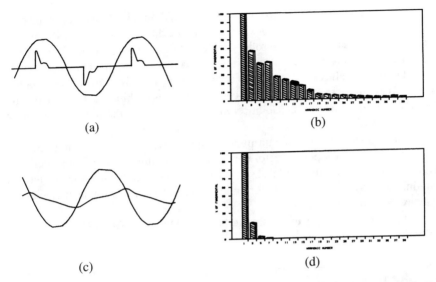

FIGURE 8-11 Comparative Analysis of CFLs (a) Voltage and current waveforms with an integral electronic ballast (b) Frequency spectrum for the CFL in Figure 8–11(a) (c) Voltage and current waveforms with an integral magnetic ballast (d) Frequency spectrum in Figure 8–11(c).

8.10 Future Developments of Electronic Ballasts

Electromagnetic ballasts have demonstrated good reliability due to their relative simplicity. Electronic ballasts, with much higher complexity and relatively fragile active semiconductors, have exhibited failure rates significantly greater than electromagnetic ballasts. As electronic lighting systems become more commonplace, reliability of electronic ballasts become more and more of an issue (Nemer 1994).

The evolution of the electronic ballast from simple invertor to the "smart ballast" of tomorrow, has meant a significant increase in circuit complexity and performance. At the same time, the end user expects a system that provides light on demand every time he or she throws a switch. Compared to many electronic devices, the ballast operates in a hostile environment as defined by ambient temperature. Excessive heat shortens component life.

Quality is essential but does not necessarily equate to reliability. There are important dependent relationships between quality and reliability that include mechanical electrical and economic considerations. Reliability can be improved at three levels. First, the use of high quality components; second, the use of high performance designs, and, third, the use of highly reliable manufacturing techniques.

With the current world energy situation, more and more electromagnetic ballasts will be replaced with electronic lighting. Thus the need for reliable electronic ballasts will keep increasing.

In the same way that electronic ballasts have dramatically increased the efficiency of light production, the next generation of dimming ballasts will provide dramatic energy savings by controlling light more effectively. Dimming electronic ballasts allow strategies, such as daylighting and compensation for lamp depreciation.

Dimming ballasts are available today, but most utilize low voltage control wiring where the cost of installing the control wiring is prohibitive. Integrated wireless control and dimming capabilities will be the basis of the next generation of "intelligent ballasts."

Also modern IC manufacturing techniques have made it possible to include complete circuitry for power factor correction and dimming control on a single IC. For example, Micro Linear's ML4830 is an IC with low distortion, high efficiency continuous boost power factor correction together with selectable variable frequency dimming and starting.

For further reading on electronic ballasts references Wood (1994) to Hagar (1993) are recommended.

References

1. Brian, B. "New Lamps for Old." *IEE Review*, November 1995, pp 239–243.
2. Mortimer, G. W. "Electronic Lighting Reliability: The Nature of the Beast." Proceedings of the HFPC/Electronic Ballast Design and Control, April 1994, pp 85–88.
3. Wood, P. N. "Hexfets Improve Efficiency, Expand Life of Electronic Lighting Ballasts." Application Note AN 973, Parts 1 to 4, International Rectifier, 1988.
4. Wood, P. N. "Simplified Ballast Designs Using High Voltage MOS Gate Driver." Proceedings of the HFPC/Electronic Ballast Design and Control, April 1994, pp 49–55.
5. Datta, S. "Comparative Analysis of Fluorescent Lighting with Electronic and Magnetic Ballasts." Proceedings of the HFPC/Electronic Ballast Design and Control, April 1994, pp 9–18.
6. Nemer, K. J. "One Utility's Experience with Electronic Ballast Systems." Proceedings of the HFPC/Electronic Ballast Design and Control, April 1994, pp 26–30.
7. Wood, P. N. "Electronic Ballasts Using the Cost Saving IR 2155 Driver." Application Note AN 995, International Rectifier, 1994.
8. Verderber, R. R. "Analysis of Past and Future Electronic Ballast Markets." Proceedings of the HFPC/Electronic Ballast Design and Control, April 1994, pp 2–8.
9. Pfendt, H. "A Case Study of Lighting Savings." Proceedings of the HFPC/Electronic Ballast Design and Control, April 1994, pp 25–30.
10. Lucas, J. R. and V. B. Wijekoon. "Power Factor and Harmonic Distortion of Compact Fluorescent Lamps." Transactions of the IESL, Sri Lanka, 1995, pp 29–38.
11. Hagar, J. "Electronic Ballast: Factor Family, Quality and Reliability." Proceedings of Power Quality, October 1993, pp 52–55.

CHAPTER **9**

Power Factor Correction and Harmonic Control

Co-Author: Aruna Ranaweera

9.1 Introduction

Power factor is not something which worries most electronics engineers. To them, it is something they learned one day in school as being cos ϕ. This conventional definition of power factor is only valid when considering pure sinusoidal signals for both current and voltage waveforms.

But the reality is something else, because most off-line power supplies now in use in computers and many other types of equipment draw a non-sinusoidal current. These supplies work directly from AC line, and have a typical front end section made by a rectification bridge and an input filter capacitor (Figure 9–1a).

A current flows to charge the capacitor only when the instantaneous AC voltage exceeds the voltage on the capacitor. Thus, a single phase off-line supply draws a current pulse during a small fraction of the half cycle duration. Between those current peaks, the load draws the energy stored inside the input capacitor.

This severely distorted non-sinusoidal AC input current generates input current harmonics. But since the corresponding input voltage distortion is negligibly small as can be seen from Figure 9–16, input voltage can be considered nearly sinusoidal. Therefore, there is no useful power produced by the current harmonics, and, the power required by the load is supplied only by the line frequency component of the input current.

In other words, the pulsed current waveform produces non-efficient extra rms current, reducing then the real power available from the mains. Thus the harmonics in the input current has resulted in a reduced power factor for the load.

So, the power factor is much more than simply cos ϕ in this case, compared with the conventional purely sinusoidal waveform case, in which poor power factor is caused by large ϕ, the phase angle difference between the voltage and current waveforms. Therefore, the conventional solution to improve power factor—addition

244   POWER ELECTRONICS DESIGN HANDBOOK

(a)

(b)

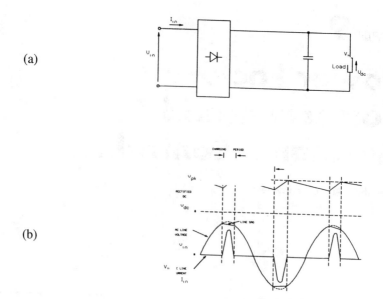

FIGURE 9-1   Full wave bridge rectifier and waveforms (a) Circuit (b) Waveforms

of appropriate value of capacitance to reduce large φ caused by the inductance of the load—would no longer help.

Here, improving the poor power factor caused by current harmonics in fact means eliminating those harmonics and making the input current as sinusoidal as possible, which is known as "current shaping."

There are two major reasons for the recent surge in interest in power factor correction. The first is that equipment with higher power rating may be operated from the power line when the power factors of the equipment are high, and the second reason is that the IEC standards force manufacturers to apply power factor correction to most equipment. This chapter provides an overview of power factor correction and harmonic control as applied to electronic products and power electronic subsystems.

9.2   Definitions

9.2.1   Power Factor

The power factor is defined by:

$$\text{power factor} = \text{Real Power} / \text{Total Apparent Power}$$
$$= P/(V*I) \quad (9.1)$$

where V and I are the total rms values of the input voltage and current, respectively. The apparent power is expressed in volt-amperes, while the real power is expressed in watts.

9.2.2 Total Harmonic Distortion

A harmonic is a component of a periodic wave having a frequency that is an integral multiple of the fundamental power line frequency. For example, Figure 9–2(a) shows a 50 Hz waveform and a 250Hz (50Hz x 5), 5th order harmonic of the fundamental frequency. Figure 9–2(b) shows the corresponding waveform when the fundamental and the 5th order harmonic are combined. The resulting waveform which is not a sinusoid is a waveform with harmonic distortion. Harmonics are a steady state phenomenon, and should not be confused with the power transients which also distort the waveforms for a very small period of time.

The level of voltage or current harmonic distortion existing at any one point in a power system can be expressed as the Total Harmonic Distortion (THD) of the current or voltage waveform. The THD for a voltage waveform is given by the following formula:

$$V_{THD} = (V_2^2 + V_3^2 + \ldots + V_N^2)^{\frac{1}{2}} / V_1 \tag{9.2}$$

where V_1 is the rms value of the fundamental voltage, and V_n for n = 2, 3, 4, etc. are rms values of the respective harmonic voltages.

9.3 Harmonics and Power Factor

Harmonic distortions are usually caused by the use of nonlinear loads by the consumers of electricity. Nonlinear loads, a vast majority of which are loads with power electronic devices, draw current in a non sinusoidal manner, and with the increased use of such devices in consumer loads, the presence of distortions in current and voltage waveforms have become a frequent occurrence today.

For instance, virtually all modern computers incorporate an SMPS for converting mains AC voltage to DC voltages required by the computer circuits. The conventional SMPS is basically an AC-to-DC converter operating from a bulk storage capacitor which is recharged directly from the line through the input rectifiers. Since the rectifiers can only conduct when the line voltage is higher than the capacitor voltage, the charging of the capacitors takes place only during a small time period, resulting in narrow input current pulses as shown in Figure 9–1(b).

Although the input current is nearly in phase with the voltage, the narrow pulses of the current have a peak value many times higher than would be present with a sinusoidal current waveform. The voltage waveform too is distorted, though much less than the current, with the peak flattened due to voltage drops caused by the high peak current flowing through the line impedance.

Only voltages and currents of the same frequency can produce non zero active power. Also, sometimes it may be reasonable to assume that the voltage is purely sinusoidal, despite the distortions in the current, and under such an assumption, V, the rms value of the voltage is equal to the rms of the fundamental frequency voltage V_1 and:

$$P = V_1 I_1 \cos \phi_1 = V I_1 \cos \phi_1 \tag{9.3}$$

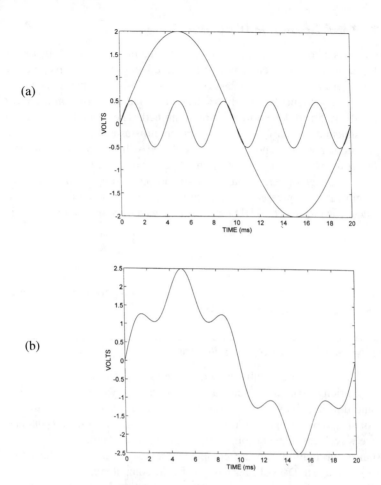

FIGURE 9-2 Fundamental and the 5th order harmonic of waveforms (a) Individual waveforms (b) Combined waveform

$$\text{Power factor} = VI_1 \cos \phi_1 / VI = I_1 \cos \phi_1 / I \qquad (9.4)$$
$$= \text{In phase fundamental rms current / Total rms current}$$

An important observation that can be made here is that, even when the fundamental current is in phase with the voltage, i.e., $\cos \phi_1 = 1$, the power factor can be less than one, since I_1 is less than I when distortions are present in the current waveform. Conversely, as the harmonic content of the current approaches zero, the power factor approaches unity, provided there is no phase lead or lag.

The typical, non corrected, off-line power supply has a power factor between 0.6 and 0.7, a value that many suppliers and users have accepted in the past, believing that any attempt to improve it would add excessive cost to the power supply. Adjustable Speed Drives (ASDs), Uninterruptible Power Supplies (UPS), electronic ballasts in energy saving lamps, and arc welders are some of the other common non-linear loads which draw current with harmonic distortion.

9.4 Problems Caused by Harmonics

Now let us look at the problems caused by these current harmonics to the rest of the utility system. To illustrate the problem due to current harmonics i_h in input current i_s of a power electronic load, consider the simple block diagram in Figure 9–3.

Due to the non-zero internal impedance of the utility source L_s (including the line connecting the source to the load), the voltage waveform at the point of common coupling to the other loads v will be:

$$v = v_s - i_s L_s \qquad (9.5)$$

Since i_s contains harmonics i_h, v the voltage at the point of common coupling is distorted, and thus the supply voltage to the other loads, which may cause them to malfunction.

On the other hand, the low value of load power factor due to the harmonics means larger currents drawn from the utility, or higher VA consumption, for a given real load (in watts). This increases the burden on the utility by overloading the system and consequent need for increased system capacity.

As an example, consider the power available to a DC load of a piece of equipment using two types of front end AC to DC conversions. The first uses a conventional rectifier and capacitive filter/storage element. The second uses a power factor corrected circuit.

On a conventional 15 A, 120 V circuit, the continuous rms current must stay below 12 A to comply with the UL limits for a 15 A circuit breaker. Accordingly, only 1440 VA are available. When the typical converter efficiency of about 75 percent and the power factor of 0.65 are considered, a conventional power supply can deliver only 702 W (1440 x 0.65 x 0.75) to the load. A power factor corrected front end, with an efficiency of 95 percent, can improve the power factor to 0.99, and, as a result, the power available to the load is 1015 W (1440 x 0.99 x 0.95 x 0.75). The

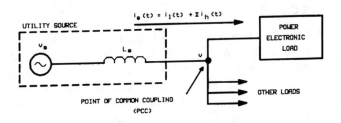

FIGURE 9–3 Utility-user interface

extra power available is significant because of the ever-increasing power demands of today's computer equipment.

In addition to the issues discussed so far, some other problems caused by harmonic currents are: additional heating and possibly over-voltages due to resonance conditions in the utility's distribution and transmission equipment, errors in metering and malfunctioning of utility relays, interference with communication and control signals, and so on. Some consequences of harmonics include unexpected failure of motor drives and power supplies, and random tripping of circuit breakers. Frequently harmonic distortion results in down time and productivity loss.

9.5 Harmonic Standards

In view of the widespread use of power electronic equipment connected to utility systems, various national and international agencies have proposed limits on harmonic current injection to the system by this equipment, to maintain good power quality. As a result, the following standards and guidelines have been established that specify limits on the magnitudes of harmonic currents and harmonic voltage distortion at various harmonic frequencies. (*Power Quality Assurance Journal*, September/October (1995) and Mohan, Undeland, and Robbins (1989)).

(a) IEC Harmonic Standard 555-2, prepared by the International Electrotechnical Commission, and accepted by its National Committees of the following countries: Austria, Australia, Belgium, Canada, Egypt, Finland, France, Germany, Hungary, Ireland, Japan, Korea (Republic of), Netherlands, Norway, Poland, Romania, South Africa (Republic of), Switzerland, Turkey, and the United Kingdom.

(b) EN 60555-2, "The Limitation of Disturbances in Electricity Supply Networks caused by Domestic and Similar Appliances Equipped with Electronic Devices," European Norm prepared by Comite Europeen de Normalisation Electrotechnique, CENELEC. This in fact is an adoption of the IEC 555-2 standard to the CENELEC member countries, and will be same for all countries in the European Union (Cogger, Singh, Hemphill 1994).

(c) IEEE Standard 519-1992, "IEEE Recommended Practices and Requirements for Harmonic Control in Electrical Power Systems."

9.5.1 IEC 555-2

IEC 555-2 provides harmonic current limits for all electrical and electronic equipment having an input current up to 16 A, intended to be connected to public distribution systems of nominal 50 or 60 Hz frequency. The voltages covered are 220-240 V single phase, and 380-415 V three phase. The equipment is classified into four groups: "Class A, for balanced 3-phase equipment and anything else that does not fit into another group; Class B, for portable tools; Class C, for lighting equipment; and Class D, for equipment having an input current with a special waveshape." (IEC Standard Publication 555-2 1982).

The harmonic limits are defined in absolute values, irrespective of the equipment's power rating. Table 9–1 below gives the maximum permissible harmonic current in amperes in these four groups according to the IEC 555-2. The same limits are shown in a bar chart in Figure 9–4.

TABLE 9–1 IEC 555-2 Harmonic Current Content Limits

	Odd Harmonics						Even Harmonics			
Harmonic Order (n)	3	5	7	9	11	13	15<n< 39	2	4	6	8 < n < 40
Maximum Permissible Harmonic Current (A)	2.30	1.14	0.77	0.40	0.33	0.21	0.15 x 15/n	1.08	0.43	0.30	0.23 x 8/n

9.6 Power Factor Correction

Power factor correction or control in these situations, where low value of power factor is a result of the presence of distortions in the current waveform, fundamentally means eliminating those distortions. In other words, the goal is to make input current look like input voltage on a moment-by-moment basis, the same as it would with a resistive load. Figure 9–5 shows line voltage and line current with and without power factor correction.

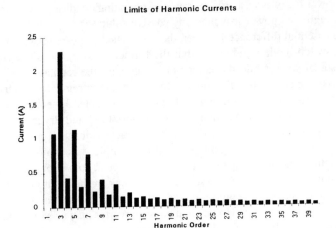

FIGURE 9–4 Limits of harmonic current content according to IEC 555-2

250 POWER ELECTRONICS DESIGN HANDBOOK

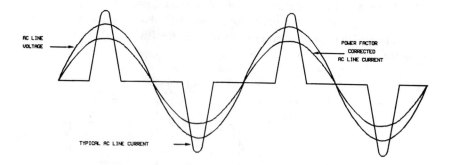

FIGURE 9-5 Effect of power factor correction on line current

In low power supplies, this can sometimes be accomplished by adding an inductor in front of the bulk capacitor. At high power levels, a more sophisticated passive component solution, such as a ferroresonant input transformer or tuned input filter may be possible. While these approaches may be able to improve power factor to greater than 0.9 in power systems, they all suffer from the growing bulk, weight, and cost as power levels increase. In addition, they are only effective over a limited range of operating voltages and power levels.

9.6.1 Active Power Factor Correction

Most designers now seem to agree that a more effective solution can be achieved with the active approach which incorporates a separate switch mode power converter configured as a current-programming pre-regulator. While adding complexity to the power supply, this type of pre-regulator can be implemented to yield a higher power factor over a broader range of operating conditions with smaller size and less weight and power loss than any passive solution.

The essential difference between the power factor pre-regulator and the conventional switch-mode regulator is that the former restores the input current to a near-sinusoidal state, while the latter only deals with the regulation of the output voltage. Like the conventional regulator, the power factor regulator can use any one of the three basic converter topologies described in Chapter 3, namely: buck, boost, and buck-boost (fly-back) (Mammano and Neidorf 1994 and Pryce 1991). Each of these topologies (Figure 9-6) has distinctive characteristics.

The buck converter (Figure 9-6a), for example, has major limitations. In order to regulate the output, the output voltage of a buck converter must be less than its minimum input voltage. Because there is a break in the input current when the input voltage falls below the output voltage, the buck converter cannot provide optimal power factor correction. However, a buck converter may be satisfactory for low output voltage applications having moderate power factor requirements. The chopped current of the buck regulator, which can generate considerable line noise

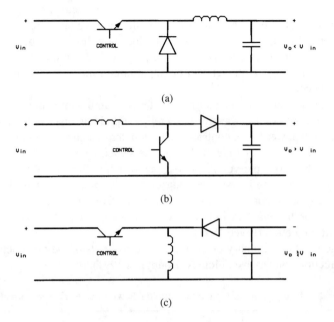

FIGURE 9-6 Three basic converter topologies that can be used for power factor correction (a) Buck converter (b) Boost converter (c) Buck-boost converter

that is difficult to filter, is also a major concern. Another disadvantage is that the maximum input voltage appears across the switch, and its base (gate) drive usually requires level shifting to a floating reference. However, because the switch is at the input, the buck converter can control input surge current and also provide protection against an output overload or short circuit.

The buck-boost (fly-back) converter (Figure 9-6c) can either step down or step up the input voltage. Because of this, it offers the advantage of an independent choice of output voltage. Also, it is simple to isolate the output (with an extra winding on the inductor), so that a single stage can accomplish power factor correction and isolation. In a circuit that is not transformer coupled, a buck-boost converter inverts the output polarity with respect to the input. Since in this topology there is no path to AC ground unless the switch is closed, the buck-boost converter can provide current limiting, during both steady state operation and initial turn-on.

One disadvantage of the buck-boost converter is that the switch has to withstand the sum of the input and output voltages. Also, the buck-boost converter operates in the discontinuous mode, which causes it to exhibit peak currents two to four times higher than the average line current. The noise created by this can be a serious problem. Because these high peak currents degrade efficiency too (especially at higher power levels), power factor correction circuits using this topology are limited to about 150W in practical applications. Fluorescent lamp ballasts and personal computer power supplies are typical applications.

Probably the most popular topology for a power factor pre-regulator is the boost converter (Figure 9–6b), which steps up the input voltage. Because the boost converter operating in the continuous mode does not chop the input current, and because the inductor itself acts as a line filter, RFI and EMI problems are generally reduced. Also, having the inductor in the input circuit makes it easy to implement current mode control.

Another advantage of the boost converter is its ability to maintain control over the complete input voltage waveform, thereby minimizing distortion, an important consideration in power factor control. In addition, the switch's common emitter configuration makes it easy to drive the base of the switch with ground referenced control signals. Moreover, the voltage across the switch is limited to the value of the output voltage. Because its peak current tends to be much less than that in other topologies, the boost converter is particularly effective for use in high power supplies. The major disadvantage of the boost converter is its inability to easily provide short circuit protection (Pryce 1991).

Table 9–2 is a summary of the merits and demerits of the three active power factor correction converter topologies (Bourgeois 1991).

TABLE 9–2 Comparison of Converter Topologies with Power Factor Correction

Boost	Buck	Buck-boost
High PF correction	Poor PF correction	High PF correction
$V_{out} > V_{in}$	$V_{out} < V_{in}$	V_{out} free
Small filter required	Large filter required	Large filter required
Switching voltage rating V_{out}	Switching voltage rating V_{in}	Switching voltage rating nV_o+V_{in} (Note 1)
	Inrush current protection	Inrush current protection
	Short circuit protection	Short circuit protection
	Floating driver	Output isolation

Note 1: n is the transformer ratio.

9.6.2 Implementation Using Boost Topology

The principle of operation of active power factor correction circuits is described here using the boost topology. The boost converter is generally preferred over buck or buck-boost converters for current shaping because in most applications, among other things, it is desirable to stabilize the DC output voltage V_o slightly in excess of the peak of the AC input voltage v_s.

This converter is shown in Figure 9–7(a) where C_o is used to minimize the ripple in V_o and to meet the energy storage requirement of the power electronic system. The DC current I_{load} represents the power supplied to the rest of the system and the high frequency component in the output current is effectively filtered out by C_o.

Power Factor Correction and Harmonic Control 253

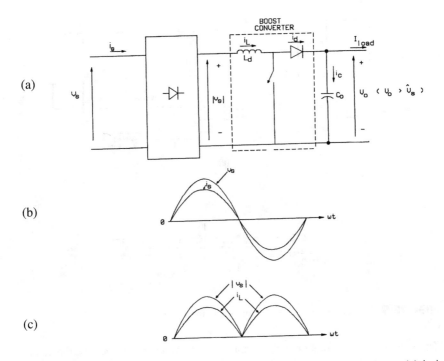

FIGURE 9-7 Active current shaping (a) Boost converter (b) Line waveforms (c) $|v_s|$ and i_L

The basic principle of operation is straightforward. At the utility input, the current i_s is desired to be sinusoidal and in phase ($\phi = 0$) with v_s as shown in Figure 9-7(b), so that the power factor ($\cos \phi$) is equal to 1. This means that, at the full bridge rectifier output in Figure 9-7(a), its output current i_L has to be of the same shape as the output voltage, which is the fullwave rectification of the supply voltage or $|v_s|$, as shown in Figure 9-7(c). In other words, i_L has to follow $|v_s|$, so that i_s will follow v_s.

However, i_L is also the input current to the boost converter, and by operating the converter in the current regulated mode, i_L can be shaped according to the supply voltage v_s as desired. The feedback control is shown in a block diagram in Figure 9-8(a). First, the output voltage V_o of the converter is sensed and it is compared with the desired output voltage V_{ref}. The difference is fed to a PI regulator, the output of which is the error signal to the multiplier (Figure 9-8(b)). Then i_M, the desired value of the current i_L is obtained by the multiplication of this error signal with the desired shape $|v_s|$, and therefore, i_M has the same shape as $|v_s|$. The status of the switch in the boost converter is now controlled by comparing the actual current i_L with the desired current i_M.

Once i_L and i_M are available, there are various ways to implement the current mode control of the boost converter, as discussed in Chapter 3. Among them, the fixed frequency pulse width modulation (PWM) is the most widely used technique.

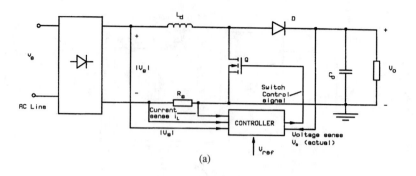

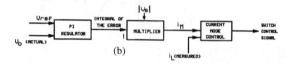

FIGURE 9-8 Control of power factor correction circuit

Thus, the switching frequency f_s is kept constant. When i_L reaches i_M, the switch in the boost converter is turned OFF. The switch is turned ON by a clock at a fixed frequency f_s, which results in i_L as shown in Figure 9–9 (Mohan, Undeland, Robbins 1989). As can be seen, the average value of i_L has the desired shape which is $|v_s|$.

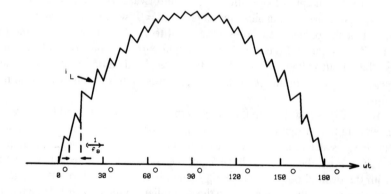

FIGURE 9-9 Inductor current i_L of the boost converter with current mode control

9.6.2.1 Output Capacitor and Boost Inductor

The parameters that affect the choice of the output capacitance C_o, which is used to minimize the ripple in V_o and to meet the energy storage requirement of the power electronic system, are:

(a) Hold-up time capability, usually 20 ms for computer power supplies.
(b) Ripple current handling capability.
(c) Allowable third harmonic distortion.

Its value can be calculated using the formula,

$$C_o = 2P_o t_{hld} / (V_{o,min}^2 - V_{oL,min}^2) \tag{9.6}$$

where, C_o = output capacitance, P_o = output power of the converter, t_{hld} = hold-up time, normally 20 ms, $V_{o,min}$ = minimum value of the output regulated voltage, and $V_{oL,min}$ = minimum input voltage of the driven load, usually a switching power supply.

The value of the boost inductor affects many other design parameters. Most of the current that flows through this inductor is at low frequency. This is particularly true at the lowest input voltage where the input current is the highest.

Normally the acceptable level of ripple current is between 10 and 20 percent. For a switching frequency of 100 kHz, the following formula will produce acceptable results.

$$L_d = 300 / P_o \text{ mH} \tag{9.7}$$

A good detailed description of the calculation procedure can be found in Micro Linear Application Handbook (1995).

9.6.3 Implementation Using Buck-Boost Topology

Unlike the boost topology, typical buck-boost topology implementation does not require a multiplier. This makes the buck-boost implementation much simpler, but it can be used only in low power applications, typically abut 150 W or below.

In a typical application, the buck-boost regulator operates in the discontinuous inductor current conduction mode. The inductor energy stored during the "ON" time of the switch is completely delivered to the output capacitance during the "OFF" time (Figure 9–6c).

At the beginning of each "ON" time the inductor current starts to ramp up from zero to a value that is proportional to the instantaneous value of the supply voltage as shown in Figure 9–10.

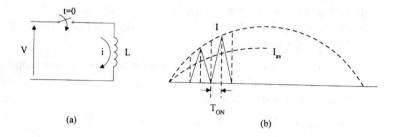

FIGURE 9-10 Principle of the buck-boost topology used in power factor correction

The instantaneous voltage:

$$v = L\frac{di}{dt}$$

$$\frac{di}{dt} = \frac{v}{L}$$

Assume the voltage v is constant and equal to $\pm|v|\pm$ during the small "ON" time period of the switch T_{ON}. Thus the current will reach a maximum value I according to the equation:

$$\int_0^I di = \frac{|v|}{L} \int_0^{T_{ON}} dt \qquad (9.8)$$

$$I = \frac{|v|}{L} T_{ON} = \frac{T_{ON}}{L}|v| \qquad (9.9)$$

$$I \propto |v| \qquad (9.10)$$

Thus the current at the end of each "ON" time is proportional to the instantaneous voltage at that time, and the average current I_{av} is a sinusoid as shown in Figure 9-10b and 9-16.

9.6.4 Multi-Stage Off-Line Converter

The obvious and most commonly used solution for providing PFC in a real-world system which requires both isolation and lower output voltage levels is a two-stage converter topology as shown in Figure 9-11. While this configuration could be implemented in many different forms and with many different control elements, what has been shown here is the most basic approach which would be applicable to medium or lower power levels where a single-ended forward topology would be appropriate for the down converter. By providing separate controls for the PFC and

Power Factor Correction and Harmonic Control 257

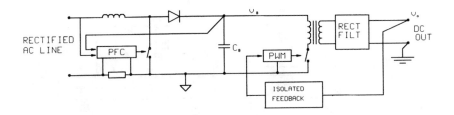

FIGURE 9-11 Two stage converter

buck stages, mixing and matching can be done to optimize the designs for each stage to meet performance and cost goals for a specific application (Mammano 1996).

A two-stage converter will use a boost circuit for PFC followed by a buck stage to lower the output voltage and incorporate line isolation.

9.7 Power Factor Correction ICs

Active power factor correction requires special control circuits that are able to force the input current waveshape to be sinusoidal and in phase with the input sinusoidal voltage, as explained in the previous section. Traditionally, a decision to incorporate an active high power factor pre-regulator would not be taken lightly, particularly at high power levels, because of the complexity of circuitry required. However, recognizing the existing and impending needs to provide a practical solution to the problem of active power factor correction, several vendors of integrated circuits have introduced devices specifically dedicated to the task. Among these vendors are Unitrode, Micro Linear, Motorola, Samsung, Cherry Semiconductor, and Silicon General.

9.7.1 UC3854

Unitrode's UC3854 is used in a boost topology and the chip includes a voltage amplifier, a low-offset analog multiplier, a current amplifier, a fixed frequency PWM, an oscillator, and a 1A totem-pole MOSFET gate driver. Also included is an enable comparator, and an over-current comparator, in addition to the 7.5V voltage reference. The greatly simplified circuit in Figure 9-12 illustrates the basic operation of UC3854.

As shown, the circuit uses a current loop as well as a voltage loop for control. The current loop samples the output waveform of the bridge rectifier through resistor R_7, which converts the voltage to a current waveform at pin 6 of UC3854. This input acts as a reference for the multiplier. To implement feed-forward line regulation, resistors R_6, R_8 and R_9, together with capacitors C_3 and C_4, develop a DC voltage at pin 8 that is proportional to the rms value of the line voltage.

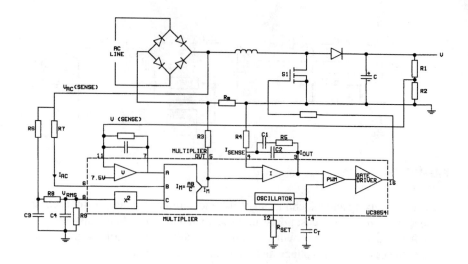

FIGURE 9-12 Power factor correction circuit using UC3854 (Reproduced by Permission of Unitrode Inc., USA)

The chip squares this DC voltage and applies it to the multiplier. To complete the voltage loop, resistors R_1 and R_2 provide a sample of the output voltage V_o and apply it to the input of the voltage amplifier at pin 11. The output of the voltage amplifier goes to the multiplier.

The output of the multiplier is a current that is equal to the product of the voltage amplifier output and the input line voltage ($|v_s|$), divided by the square of the rms line voltage ($I_M = AB/C$). Acting as a control signal to the pre-regulator, I_M has an instantaneous value that follows the shape of the input voltage, and an average value that is inversely proportional to its rms value. Resistor R_3 converts I_M to a voltage while a voltage proportional to the line current is measured across R_s, the line current sensing resistor.

The difference between these two voltages is applied to the input of the current amplifier. Under closed loop control, the current amplifier will try to keep this voltage difference close to zero volts. This forces the voltage produced by the return current on R_s to be equal to the voltage across R_3.

The amplified current error signal is then applied to the inverting input of the PWM comparator. The other input of the PWM comparator is the ramp generated by the timing capacitor C_T of the oscillator. Pulse width modulation is obtained when the amplified error signal that sets up the trip point modulates up and down.

The output of the PWM is connected to a totem pole driver capable of sourcing and sinking ±1 A peak. As shown in Figure 9-12 the PWM and the gate driver control the ON-OFF action of the MOSFET power switch S_1 to force the input line current to follow the programmed value. The UC3854 uses average current control; the compensation network comprising R_4, R_5, C_1, and C_2 performs the averaging.

Power Factor Correction and Harmonic Control

R_{SET} programs the charging current of the oscillator and the maximum output of the multiplier. C_T sets the PWM oscillator frequency.

The maximum output voltage level is internally limited to 18V, while the under-off voltage lockout comparator incorporated with the output will turn the circuit off before the PWM output drops below 8V. Since the chip uses average current control, it can accurately maintain sinusoidal line current without resorting to slope compensation (Mammano and Neidorf 1994 and Pryce 1991).

UC3854 provides power factors to 0.99, and limits the line current distortion to less than 5 percent. The chip operates in systems with line voltages of 75 to 275 V and line frequencies of 50 to 400 Hz. When compared with other power factor controllers, the UC3854's higher reference voltage (7.5V) and higher oscillator output (5V) can offer advantages in high-power supplies, which typically exhibit high noise.

9.7.2 ML Series of ICs

Micro Linear Corporation offers several power factor correction ICs. Among them, ML4812 is designed to optimally facilitate a boost type power factor correction system. Special care has been taken in the design of the ML4812 to increase system noise immunity. The circuit includes a precision reference voltage, oscillator, multiplier, error amplifier, over-voltage protection, ramp compensation as well as high current output. In addition, start-up is simplified by an under-voltage lock-out circuit with 6V hysteresis. A block diagram of ML4812 showing the basic building blocks is given in Figure 9–13.

The operation of the circuit is similar to that of UC3854, as explained before. A fraction of the output voltage is fed to the error amplifier through pin 4. A current

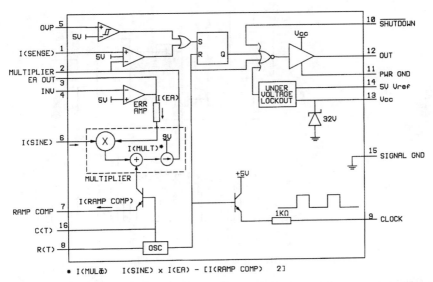

FIGURE 9–13 Functional block diagram of ML4812 (Reproduced by permission of Micro Linear Corp., USA)

that is proportional to the input fullwave rectified voltage is fed to the pin 6. The output of the multiplier is applied to a resistor in the external circuit through pin 2. The difference between the voltage produced by this and a voltage proportional to the line current which is fed to pin 1, is applied to the PWM comparator which in turn controls the switch.

In a typical application, the ML4812 functions as a current mode regulator. The current which is necessary to terminate the cycle is a product of the sinusoidal line voltage times the output of the error amplifier which is regulating the output DC voltage. Ramp compensation is programmable with an external resistor to provide stable operation when the duty cycle exceeds 50 percent. The use of ML4812 in a continuous mode boost circuit using current mode control scheme is shown in Figure 9–14 with the resulting current waveforms. Typical applications include computer systems that require optimum power factor correction.

ML4813 is designed to optimally facilitate a discontinuous buck-boost type power factor correction system for low power, low cost applications. A functional block diagram of ML4813 is shown in Figure 9–15. Similar to ML4812, this too has a precision reference voltage, oscillator, error amplifier, high current output, and an over-voltage comparator, and special care has been taken in the design to reject system noise. This too has an under-voltage lockout circuit with 6V hysteresis.

However, unlike ML4812, ML4813 does not have a multiplier and there is no multiplication of the error amplifier output with the input fullwave rectified voltage $|v_s|$. Instead, in ML4813, "ON" time of the switch is set by the error amplifier and the PWM comparator according to the instantaneous value of $|v_s|$, a portion of which in this case is the external input to the error amplifier.

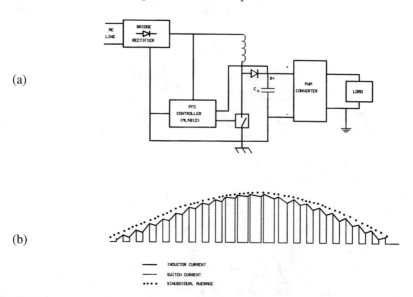

FIGURE 9–14 Boost mode power factor correction using ML4812 and inductor current waveform (Reproduced by permission of Micro Linear Corp., USA) (a) Block diagram (b) Waveforms

Power Factor Correction and Harmonic Control 261

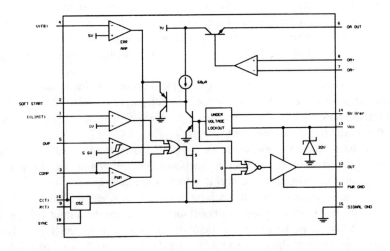

FIGURE 9-15 Functional block diagram of ML4813 (Reproduced by permission of Micro Linear Corp., USA)

In a typical application, the ML4813 functions as a voltage mode, discontinuous current, buck-boost power factor regulator as shown in Figure 9-16. The regulator operates in the discontinuous inductor current conduction mode. The inductor energy stored during the "ON" time of the power switch Q is completely delivered

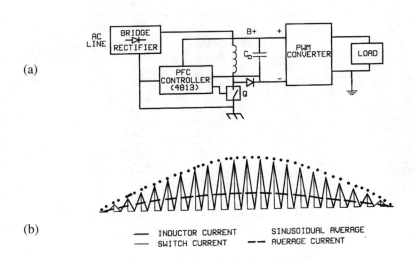

FIGURE 9-16 Buck-boost power factor correction using ML4813 and inductor current waveform (Reproduced by permission of Micro Linear Corp., USA) (a) Block diagram (b) Waveforms

to the output capacitance during the "OFF" time. At steady state conditions, the inductor current at the beginning of the "ON" time starts to ramp up from 0 Amps to a value that is determined by the instantaneous value of the input full wave rectified voltage; the "ON" time as it is set by the error amplifier and the PWM comparator; and finally by the inductor itself.

By maintaining a constant duty cycle, the current follows the input voltage, making the impedance of the entire circuit appear purely resistive. With the buck-boost circuit, power factors of 0.99 are easily achievable with a small output inductor and a minimum of external components.

The discontinuous mode, buck-boost topology used in ML4813 is particularly well suited for low power applications such as fluorescent ballasts and low power switching supplies. Also, it is a useful topology when there is a requirement for the output voltage to be lower than the peak input voltage, or where an isolated output is required. This is not possible with the boost topology, where the output voltage must always be higher than the maximum peak of the input voltage range. The typical input range for the buck-boost power factor regulator is from 90 VAC to 260 VAC.

A comparison of the two circuits based on ML4812 and ML4813 is given in Table 9–3.

TABLE 9–3 Comparison of the two power factor correction circuits in Figures 9–14 and 9.16.

	Continuous, Boost	*Discontinuous, Buck-Boost*
Output Voltage	$V_{out} > V_{in}$	Independent of V_{in}
Input Current	Continuous	Discontinuous
Output Current	Discontinuous	Discontinuous
Control	Simple Current Mode	Simple Voltage Mode
Peak Current (150 W)	2A	9A
Transformer Isolation	Not Possible	Easy
Rectifier Needs	Fast	Moderate
V_{max} on power Switch	V_{out}	$V_{out} + V_{in}$ (peak)
Surge Current Limit	Difficult	Inherent
Input Transient Absorption	Inherent	Not inherent
Input Line Filter	Minimal	Complex π network
Usable Power Levels	75 W to Over 2000 W	Below 150 W
Controller	ML4812	ML4813

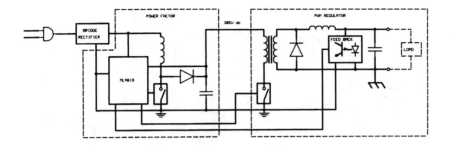

FIGURE 9-17   Combined PWM and power factor controller in ML4819 (Reproduced by permission of Micro Linear Corp. USA)

The third IC offered by Micro Linear is the ML4819, which combines a boost-mode power factor circuit similar to the ML4812 with a conventional PWM controller circuit. The PWM section can be used for either current or voltage-mode control for a second-stage converter. Figure 9-17 is a simplified diagram of the individual functions. Combining the two circuits in a single device minimises the component count and saves space. Because both circuits share the same oscillator, synchronization is inherent. Moreover, a large oscillator amplitude of 4.3 V maximizes noise immunity.

The power factor section uses peak current sensing, and the programmable slope compensation is common to both sections. The PWM section includes cycle-by-cycle current limiting as well as duty cycle limiting (for single-ended converters). Both sections feature individual 1A totem-pole output driver but the under-voltage lockout function is shared (Pryce 1991 and Alberkrack and Barrow 1993).

In addition, Micro Linear offers ML4821, the operation of which is similar to ML4812 but for power ranges above 500 W. Also, Micro Linear offers ML4824, the operation of which is similar to ML4819 but for power ranges above 1000 W.

9.7.3   MC34262, MC33262

Motorola's MC34262 and MC33262 are high performance, critical conduction, current mode power factor controllers specifically designed for use in off-line active pre-converters. These devices provide the necessary features required to enhance poor power factor loads by keeping the AC line current sinusoidal and in phase with the line voltage.

The two chips contain many building blocks and protection features employed in modern high performance current mode power supply controllers, as shown in Figure 9-18. However, unlike the previously described chips, these devices do not contain an oscillator.

264 POWER ELECTRONICS DESIGN HANDBOOK

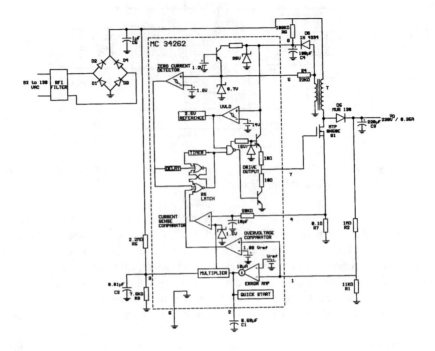

FIGURE 9-18 Power factor correction circuit using MC34262 (Copyright of Motorola, used by permission)

These devices contain an error amplifier with access to the inverting input and output. The amplifier is a transconductance type. It has high output impedance with controlled voltage to current gain. The output of the error amplifier connects internally to the multiplier, with pin 2 available for external loop compensation. The error amplifier monitors the average output voltage of the converter over several line cycles, resulting in a fixed drive output on time.

Another key feature in using a transconductance amplifier is that the input can move independently with respect to the output, because the compensation capacitor connects to ground. This allows dual usage of the voltage feedback input pin by the error amplifier and the over-voltage comparator.

The MC34262 operates as a critical conduction current mode controller, whereby the zero current detector initiates output switch conduction. Switch conduction ends when the peak inductor current reaches the threshold level established by the multiplier output. The zero current detector initiates the next ON time by setting the RS latch at the instant the inductor current reaches zero. This critical conduction mode of operation has two significant benefits.

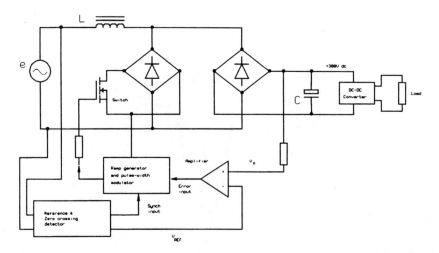

FIGURE 9-19 Active low frequency power factor correction circuit (Reproduced by permission of Unipower Corp., USA)

First, because the MOSFET cannot turn on until the inductor current reaches zero, output rectifier reverse recovery time is less critical, allowing use of an inexpensive rectifier. Second, there are no dead time gaps between cycles, so the AC line current is continuous, thus limiting the peak switch to twice the average input current. The under-voltage lockout comparator in the chip guarantees that the IC is fully functional before enabling the output stage (Alberkrack and Barrow 1993).

9.8 Active Low Frequency Power Factor Correction

All of the circuits we have discussed so far were active high frequency power factor correction circuits which typically operate at switching frequencies of 20 kHz to 100 kHz, far higher than the line frequency.

Active low-frequency (LF) PFC circuits operate at a switching frequency of twice the line frequency, 100 or 120 Hz, and in synchronism with it. This method uses an active switch, a LF inductor, and a control circuit to perform PFC.

One way to implement active-LF correction is through a patented technique used by Unipower Corp. (Figure 9–19). This circuit was chosen because of its simplicity and because it provides the best overall performance with minimum impact on size, efficiency, reliability, EMI, and cost.

The MOSFET-switch circuit is driven into conduction at the zero crossing of the AC-source voltage, and current builds up in the inductor, L, from the AC source. At a controlled time after the zero crossing (typically 1 to 2 ms), the switch is turned

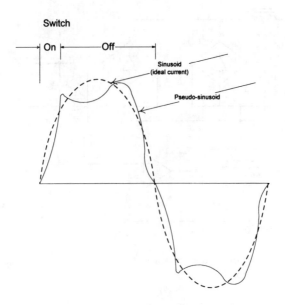

FIGURE 9-20 Pseudo-sinusoidal input current produced by active low frequency power factor correction

off and the source current through L transfers to the output rectifier and filter circuit. The MOSFET switch is driven by a pulse-width-modulation (PWM) circuit that compares an error-feedback signal to a ramp signal that is synchronized by a zero-crossing signal.

The reference for the error amplifier is proportional to the rms value of the AC-line voltage, and the output voltage (about 300 V DC) is load-regulated by adjusting the conduction time of the MOSFET switch. The input-current waveform is enhanced by maintaining an optimum relationship between the magnitude of the AC-input voltage and the DC-output voltage.

The result is a pseudo-sinusoidal current waveform which gives a high power factor of 0.96 to 0.98 with low harmonic currents (Figure 9–20).

Power factor at rated load is relatively constant over line voltage ranges of 90 to 132 V AC or 180 to 264 V AC. These input ranges can be made switch-selectable, or autoranging techniques can be used to provide automatic switching.

This method is highly efficient, with lower total losses than the HF method. Its low component count also results in high MTBF. A summary of the advantages and disadvantages of the active LF method is given in Table 9–4 (Hunter 1992).

TABLE 9-4 Active Low-Frequency Power-Factor Correction

Advantages	Disadvantages
• High efficiency	• Hold-up time changes with line voltage
• Simple circuitry	• DC bus not regulated
• High MTBF	• Larger and heavier than high frequency method
• High PF (0.96-0.98)	
• Low EMI	
• Low harmonic distortion	
• Wide frequency range	
• Lowest cost	
• Moderate input-voltage range (90 to 132V AC or 180 to 264V AC)	

9.9 Evaluation of Power Factor Correction Circuits

Input power factor, harmonic current content, and waveshape are all used when describing the performance of a power factor correction circuit. To measure power factor, one must have a very reliable power meter that accurately measures both apparent power and true average power. Some of the older model meters that measure power factor by determining the phase angle between the current and voltage waveforms will not serve our purpose.

The true average power for distorted AC waveforms can only be determined when the current and the voltage are simultaneously sampled, then multiplied and integrated. One should be very careful about which meters are trustworthy when it comes to measuring power with non-sinusoidal and noisy waveforms.

Harmonic current content also has to be measured using a similar meter or using a spectrum analyzer. The IEC 555-2 standard requires current content conformity to the 40th harmonic.

Efficiency of the power factor correction circuit also is an important indicator of the performance of the circuit. Efficiency measurements require accurate measurement of the output average power as well as the input average power. Because the output voltage as well as the output current is DC, one might just use something like a lab bench top meter to measure the two readings. However, in order to get truly accurate efficiency measurements, the same power meter should be used to measure both the output as well as the input power.

References

1. "Harmonics-Power Quality Basics." *Power Quality Assurance Journal*, September/October 1995, p 67.
2. Mohan, N., T. M. Undeland, and W. P. Robbins. "Power Electronics, Converters, Applications and Design." John Wiley & Sons, 1989, pp 409–434.
3. Mammano, B. and B. Neidorf. "Improving Input Power Factor: A New Active Controller Simplifies the Task." *Digital Designers*, April 1994, pp 57–66.
4. Pryce, D. "Specialized ICs Correct Power Factor in Switching Supplies." *EDN*, 1991, pp 106–114.
5. Bourgeois, J. M. "Circuits for Power Factor Correction with Regards on Mains Filtering." Power Conversion Proceedings, September 1991, pp 336–344.
6. Alberkrack, J. H. and S. M. Barrow. "Power Factor Controller Minimizes External Components." *PCIM*, February 1993, pp 42–48.
7. IEC Standard Publication 555–2. "Disturbances in supply system a used by household appliances and similar electrical equipment—Part 2: Harmonics" International Electrotechnical Commission, 1982.
8. Cogger, S., B. Singh, and H. Hemphill. "European EMC Standards." *Digital Designers*, April 1994, pp 83–87.
9. Micro Linear Applications Handbook, 1995.
10. Mammano, Robert A. "New Developments in High Power Factor Circuit Topologies." HFPC Power Conversion, September 1996 Proceedings, pp 63–68.
11. Hunter, Patrick L. "Solve Switcher Problems with Power Factor Correction." *Electronic Design*, February 6, 1992, pp 67–77.

Bibliography

1. Power Factor Corrector. *Application Manual, 1st Edition*. SGS-Thomson Micro Electronics, October 1995.
2. Klein, J. and M K. Nalbant. "Power Factor Correction Incentives, Standards and Techniques." *PCIM*, June 1990.
3. Shin, D. M., R. H. Chrisitiansen, and D. J. Kim. "Boost vs Buck-Boost for Power Factor Correction." HFPC, April 1994.
4. Casicio, J. L. and M. Nalbant. "Active Power Factor Correction Using a Flyback Topology." *PCIM*, August 1990, pp 10–17.
5. Sum, K. K. "Power Factor Correction For Single-Phase Input Power Supplies." *PCIM*, December 1989, pp 18–24.
6. Nalbant, M. K and J. Klein. "Design of a 1KW Power Factor Correction Circuit." *PCIM*, July 1990, pp 17–24.
7. Todd, P. C. "Power Factor Correction Control Circuits." *PCIM*, October 1993, pp 70-79.
8. Venable, H. D. "Testing Stability of Power Factor Correction Circuits." *PCIM*, February 1994, pp 52–58.
9. Wyk, J.D.V. "Power Quality, Power Electronics and Control." The European Power Electronics Association, 1993.
10. Quek, D. and S. Yuvarajan. "Control and Protection Circuit Drives Variable Power Factor MCT Converter." *PCIM*, August 1994, pp 28–39.
11. Andreycak, B. "Auxiliary Supply Tracks Power Factor." *EDN*, June 1991, pp 20.
12. Locher, R. E. and J. Bendel. "Minimize Diode Recovery Losses and EMI in PFC Boost Converters." *PCIM*, February 1993, pp 18–23.
13. Strassberg, D. "Power-Factor-Corrected Switching Power Supplies." *EDN*, April 11, 1991, pp 90–100.
14. Deuty, S., E. Carter, and Dr. A. Salih. "GaAs Diodes Improve Power Factor Correction Boost Converter Performance." *PCIM*, January 1995, pp 8–19.

15. Joseph, L. and A. Lovrich. "Microcontroller Provides SMPS Power Factor Correction Benefits." *PCIM*, March 1995, pp 32–44.
16. Mitter, C. S. "MOSFET Provides Active Inrush Control in Portable DC Power Systems." *PCIM*, May 1995, pp 10–24.
17. Hoff, M. "Instability of Power Factor Corrected Power Supplies With Various AC Sources." Power Conversion Proceedings, October 1993, pp 1–8.
18. Venable, H. D. "Testing Power Factor Correction Circuits for Stability." Power Conversion Proceedings, October 1993, pp 9–19.
19. Dierberger, K. and D. Grafham. "Design of a 3000 Watt Single MOSFET Power Factor Correction Circuit." Power Conversion Proceedings, October 1993, pp 20–33.
20. Shin D. M., R. H. Chrisitiansen, and D. J. Kim. "Boost vs Buck-Boost for Power Factor Correction." HFPC Proceedings, April 1994, pp 77–86.
21. Wodarczyk, P. J. and J. E. Wojslawowicz. "Power Devices with Integrated Active Voltage Clamp Circuits." Power Conversion Proceedings, September 1991, pp 202–210.
22. Gruzs, T. M. "Harmonic Current Reduction Techniques For Computer Systems." Power Quality Proceedings, October 1993, pp 315–325.
23. Bourgeois, J. M. "Circuits for Power Factor Correction with Regards on Mains Filtering." 336 Power Conversion Proceedings, September 1991, pp 336–344.
24. Engler, J. "Harmonic Distortion of AC Line Power and Methods for Measuring Harmonic Currents." Power Quality Proceedings, October 1993, pp 416–425.
25. Paulakonis, J. and J. Dunn. "Power Factor Corrected Microcontrolled UPS." HFPC Proceedings, April 1994, pp 18–26.
26. Garfinkel, M. "Practical Emi Techniques." Digital Designers, April 1994, pp 74–82.
27. Cohen, I. "Evaluation and Comparison of Power Conversion Topologies." The European Power Electronics Association, 1993, pp 9–16.
28. IXYS Semiconductors. "PFC Powerstage for Boost Converters." Technical Information 35, pp 1–8.
29. IEC Standard Publication 555-1. "Disturbances in Supply Systems Caused by Household Appliances and Similar Electrical Equipment." International Electrotechnical Commission, 1982.
30. Andreycak, B. "Power Factor Correction Using the UC 3852 Controlled On-Time Zero Current Switching Techniques." pp 978–993.
31. Edison Electric Institute. "Power System Harmonics." *Power Quality Assurance*, November/December 1995, pp 26–32.
32. Cho, J. G. and G. H. Cho. "Novel Off-Line Zero-Voltage Switching PWM AC/DC Converter for Direct Conversion from AC Line to 48 VDC Bus with Power Factor Correction." *IEEE*, 1993, pp 689–695.
33. Kandianis, A. and S. N. Manias. "A Comparative Evaluation of Single-Phase SMR Converter With Active Power Factor Correction." *EDP Journal*, May 1996, pp 31–36.
34. Blake, R. F. "Sine Shaping Rectifiers." *Power Quality Assurance*, March/April 1996, pp 23–31.
35. Waller, M. "The Next Dimension in Harmonic Problems." *Power Quality Assurance*, September/October 1996, pp 20–24.
36. Leman, B. "Power Factor Correction Circuit Cuts Component Count." *PCIM*, October 1996, pp 8–18.
37. Medora, N. K., A. Kusko, and R. Blanchard. "Power Factor Correction Ics—A Topological Overview." HFPC Proceedings, May 1995, pp 465–479.
38. Leman, B. R. "Simple Power Factor Correction Circuit Has Only 17 Components." HFPC Proceedings, May 1995, pp 424–434.
39. Wuidart, L. "Understanding Power Factor." AN824/0795, pp 7–10.
40. Bourgeois, J. M. "Circuits for Power Factor Correction with Regards to Mains Filtering." AN510/0894, pp 11–19.

CHAPTER **10**

Power Integrated Circuits, Power Hybrids, and Intelligent Power Modules

10.1 Introduction

From their humble beginnings as audio amplifiers, Power Integrated Circuits (PIC) are now moving towards a more prominent role in many power electronic systems. Today an increasing number of applications make use of these devices. Usually power integrated circuits are designed for specialized applications in the lower power range. Their concept represents the natural continuation of IC evolution, realizing a complete system on one chip.

The power limit arises from the amount of heat that can be dissipated by a silicon surface of approximately one square centimeter in size. When most of the silicon real estate is devoted to power devices the chips are usually called "smart power." However, the terminology applied is more often based upon institutional rather than technical reasons.

What constitutes a power integrated circuit varies from manufacturer to manufacturer, depending on their product portfolios and point of view; that is, whether they approach the market from a discrete-transistor or an IC perspective. Some define power ICs by their functions—whether the IC actually includes the power transistor itself, others, by the IC's voltage and current levels, and still others by the IC's general involvement in controlling power.

Smart power is cost-effective today in many applications because additional functions can be combined on the chip at low cost, which would be expensive or impractical to add using discrete components. These functions include drive circuitry, thermal protection, over and under voltage protection, current limiting, and diagnostics. Many smart power products were originally developed to meet the demanding environmental and cost constraints of the automotive and consumer electronics industries. As a result, industrial applications, where numbers produced are lower, also have made use of these devices in the recent past. The technology is

272 POWER ELECTRONICS DESIGN HANDBOOK

steadily progressing towards much more complex devices such as Smart Power Hybrids, Smart Power ICs, Smart Power Microcontrollers, etc., all of which could be generally grouped as Intelligent Power Modules (IPM).

10.2 Evolution of Power Integrated Circuits

Figure 10–1 indicates how two independent paths starting from discrete transistors and simple stand alone driver ICs have progressed independently towards the levels of integration in power ICs available in the mid 90s. The figure shows two independent paths of devices, beginning with discrete power transistors and driver ICs. As integration increases, each device adds functions until the transistors and driver combine into one.

With the advent of process technologies that can integrate virtually everything but the load itself, the current capabilities of ICs for power actuation and switching seem almost limitless. The potentially integrated functions include logic and control, sometimes implemented by a fully functional microcontroller; protection; diagnostic feedback; and, finally, a power output stage. Whether you call it "smart" or "intelligent" power, the technical feasibility of combining these functions into one IC using some mixture of bipolar, CMOS, and Double-Diffused MOS (DMOS) structures is proven. The process titled BCD (bipolar, CMOS, and DMOS) or its variations are used by most component manufacturers.

In the discussion of the development of PICs, it is quite clear that drivers and discrete power transistors can produce the most flexible designs. Two developments contribute to this flexibility: the wide availability of discrete transistors, particularly power MOSFETs, and the increasing numbers and types of drivers, particularly MOSFET drivers.

Regardless of the driver you select, you can independently select a MOSFET based on the design's speed and efficiency. Vendors of these MOSFETs also continue to make dramatic improvements in the device's efficiency. The on-resistance

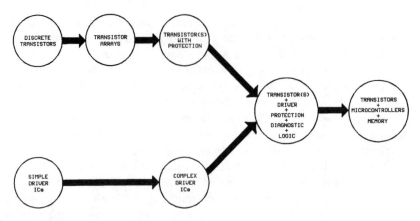

FIGURE 10–1 Progress of power integrated circuits

($r_{DS(ON)}$) of state of the art TO 220 packaged N channel MOSFETs may be around 10mΩ at 60V. A similarly packaged, state of the art P channel device has an on-resistance of around 45mΩ.

Integrated power transistors just cannot compete with discrete transistors on the basis of $r_{DS(ON)}$ alone. There is a trade-off between specific on-impedance (a measure of $r_{DS(ON)}$ versus die area) of discrete transistors and ICs. Most power ICs have higher specific on-impedances than do discrete devices, simply because of fabrication differences that result in the ICs' larger power structures. An IC requires two to four times more die area than does a discrete transistor to achieve the same on resistance; such a large die area makes ICs too expensive to produce. The on resistance values also depend on vendors' proprietary processes and voltages.

Use of discrete transistors also provides two other advantages: reduced size and increased protection. Discrete transistors are undergoing a dramatic shrinkage. Manufacturers of discrete transistors have also found inexpensive ways to add protection features, such as overcurrent protection and output-voltage clamps. In newer families such as "intelligent discretes" small transistors, diodes, and resistors etc. are added to provide intelligent device control.

These control devices can be as simple as one bipolar transistor and two resistors. Since they are fabricated within the standard power MOSFET process and generally take up a very small silicon area, the cost that they add to the base device is minimal. This is more than offset by the added value these control elements provide. Figure 10–2 indicates a very basic representative device from such a family from Harris Semiconductors Inc., which is categorized as a current limiting power MOSFET. The idea of these types of devices is to provide the designers with the basic terminal configuration (drain, source, and gate) externally, but the internal device carries many more added features to the device.

In today's component market, many types of drivers are available, particularly for driving MOSFETS and IGBTS. The term "MOSFET driver" refers to a wide range of devices. They all provide a buffer between the analog control circuitry and the true power world. These drivers suit a wide range of applications, including

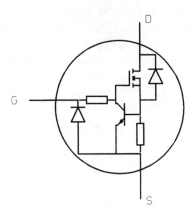

FIGURE 10–2   Current limiting MOSFET

motor controls, power supplies, UPS systems, automobile braking systems, air bag deployment, and industrial controls.

Although these drivers lack a power transistor, they can still implement protection. A separate driver can sense over current with suitable feedback. Some drivers can also sense insufficient gate drive voltage—a form of under voltage protection to prevent the MOSFET from operating in its potentially destructive linear region.

Integrated power and driver ICs can do things that drivers and discrete transistors often cannot. For example, ICs that integrate the driver and power transistor or transistors can boast about all the protection features of discrete transistors, including thermal protection, which discrete transistors cannot easily implement. ICs in particular can add thermal protection without additional components.

Combining power devices with a microcontroller and memory currently represents the highest level of integration. To date, a number of manufacturers, including Motorola, Phillips, SGS Thomson, and Texas Instruments, have demonstrated this capability (Kerridge 1994). In 1990, Motorola integrated its 8-bit 68HC05 core with 96 bytes of RAM and 2064 bytes of user ROM and 240 bytes of ROM for test functions.

More recently, SGS Thomson unveiled its third-generation BCD3 process with which the company can combine an ST6 8-bit microcontroller with a DMOS H-bridge power stage, driver and interface functions, a charge pump, and thermal protection. Using BCD3, the company can also integrate EEPROM, making the IC configurable by hardware and software. These programmable power devices or power PLDs contradict the notion that "high integration means narrow application."

10.3 BCD Technology

First introduced in 1986, mixed Bipolar-CMOS-DMOS (BCD) smart power technology has brought significant advances in the art of integrating signal and power circuits on the same chip. IC technologies combining power DMOS with signal circuits had existed previously, but BCD technology differs by the use of isolated DMOS power transistors having all of the contacts on the top surface. In contrast, other processes used discrete-type DMOS structures where the lower surface of the die is the drain contact, so that two or more DMOS devices can only be placed on the same chip if they have a common drain contact.

With the advent of BCD technology, designers were free for the first time to integrate as many DMOS power transistors as they wished, and to interconnect them in any way. Thus it became feasible to produce smart power ICs with half-bridge and bridge output stages, and even integrate several output stages on the same chip. The first commercial circuits to be produced with this technology were H-bridge motor driver ICs (operating on supply voltages of 48V and delivering 1.5 A continuous output current) and SMPS ICs capable of delivering 10A output current.

In the early 1990s the original four micron BCD technology was superseded by a shrink 2.5 micron version, bringing a doubling of both signal circuit density and current density in the power section. With the introduction of this second generation version (BCD2), designers were no longer restricted by die size concerns and quickly learned to make increasingly complex ICs. Another important trend was to develop high voltage technologies. A 250V evolution of BCD technology has been developed for applications in the industrial and telecom fields.

In the 3rd generation of BCD technology (BCD3) the lithography is reduced to 1.2 µm. Apart from the increased density, this evolution allowed the convergence of technologies too. This 1.2 µm BCD technology is compatible with the existing 1.2 µm CMOS technologies (SGS Thomson Microelectronics, 1994). CMOS transistor packing density is 5000 transistor/mm^2, three times greater than the previous generation technology. The DMOS power transistors also take up less space, due to their low resistance of 0.25 Ω/mm^2, which is half that of BCD2.

Other improvements over BCD2 include double polysilicon layers and double level, triple thickness metal layers, which help achieve the high levels of integration. The signal section density reaches a level where it is economically feasible to integrate an 8-bit microcomputer core and peripheral circuits on a smart power chip. EEPROM or EPROM memory can be added, too.

10.4 Applications of Power Integrated Circuits

To justify the distribution of non-recurrent engineering costs (involved in the development of PICs) among a large number of components, PICs are currently used in special market segments such as:

(a) Automotive applications
(b) Disk-drives
(c) PWM systems for bidirectional stepper and DCmotors
(d) Electronic ballasts for fluorescent lighting
(e) Off-the-line switchmode power supplies
(f) Special classes of power amplifiers

Typical auto applications include control of mirrors, seats, windows, and instrument panels, and all require ICs with approximately 60V, 4A rating. Many of these specialized products, particularly automotive ones, have spilled over into the general industrial market, starting out as custom ICs and eventually becoming standard offerings.

Although it seems that most highly integrated power ICs are geared toward the automotive industry, vendors do develop many ICs—specialized motor controllers and drivers, in particular for many computer peripherals and industrial applications. Some of these drivers now integrate PWM to drive bidirectional stepper and DC motors. One example is Allegro Microsystems UDN 2961B/W half bridge with 45V and 3.4A ratings (Zacher, Pyle, and Weiss 1992). This power IC is usable in printer hammer or motor driver applications.

Although the automotive and disk drive markets are huge, numerous other products target the more fragmented industrial markets. Also, emerging high volume markets are pushing the development of power ICs with high voltage and current requirements. The foremost example is the integration of high-voltage structures generally greater than 100V with logic-level circuitry. Over 20 semiconductor companies worldwide, including AT&T Micro-electronics, Harris Semiconductor, International Rectifier, Power Integrations, and SGS Thomson, are pursuing high-voltage control applications. One of the high-volume markets driving much of this work is the electronic ballast for fluorescent lighting.

Applying IC technology to electronic ballasts is hot for two reasons: a worldwide drive for energy efficiency is causing a boom in the fluorescent-lighting industry, and real money, in the form of utility-company subsidies to consumers, is behind the increased-efficient-energy effort. On the technical side, high-voltage IC technology has progressed to the point at which vendors can build low-cost ICs that can compete with existing electronic ballast designs. Older designs typically use a pulse transformer and discrete transistors.

Using an IC eliminates the need for a transformer, thus easing the design, increasing efficiency, and reducing the design's size. A recent example of a low-cost approach for this application is International Rectifier's IR 2155 IC for energy saving lamps (Swager 1994). Some application information and details about mixed technology power ICs are in Gauen (1993), Devore, Teggatz, and Compton (1994), Kanner and Wellnitz (1995), and Powers (1995).

10.5 Power Hybrids

Power hybrid ICs are available in a variety of circuit configurations, power capabilities, and packaging styles. Examples of available circuits include solid-state relays and circuit breakers, half-bridge and full-bridge drivers, 3-phase drivers, and DC/DC converters. See Table 10–1 for a representative list of power hybrid ICs.

Because of their specialized construction, hybrid ICs are often more expensive than equivalent combinations of monolithic ICs and discrete power devices. This higher cost is particularly true for military-grade hybrid ICs, which make up a large part of the market. Despite the cost, hybrid ICs have the redeeming characteristics of small size and the ability to combine specialized circuitry in a single package.

A key factor in obtaining these characteristics is the extensive use of chip-level and surface-mount components, including both active and passive types. Typically mounted on an alumina or beryllia substrate, such components minimize the size of the final package and allow flexibility in the design of the overall circuit.

A power hybrid can be as simple as a half-bridge driver or as complex as a DC/DC converter that contains all of the necessary active and passive devices, including magnetic components and filtering elements. Highly specialized hybrids often contain dozens of components and can be quite elaborate.

Another common circuit block implemented in power hybrid technology is the operational amplifiers for high voltage and high current applications. For example, PA89 (Figure 10–3) power operational amplifier from Apex Microtechnologies Inc. is capable of providing 75mA of continuous output current while operating over a supply range of ±75 to ±600V. The op amp is designed for use in specialized applications such as piezoelectric drives, high-voltage instrumentation, and electrostatic deflection. With its output-bridge configuration, the op amp could provide a 1000V p-p output signal with programmable current limiting.

For applications where a standard hybrid circuits are unable to provide customer requirements, several companies supply custom circuits. Phillips Circuit Assemblies, for example, has a number of technology choices from which they can fabricate application-specific circuits. Such companies work directly from the user's schematic and convert it to a hybrid circuit that saves space and usually has superior environmental ruggedness and greater reliability than conventional approaches. Hybrid modules with IGBTs are discussed in section 10.7.

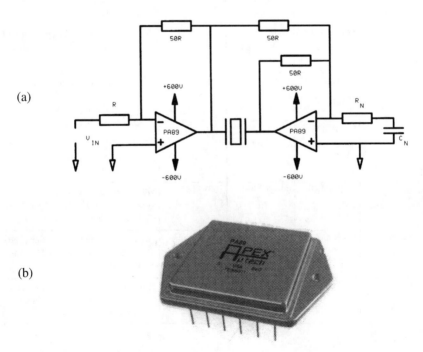

FIGURE 10–3 A power operational amplifier in hybrid form used for a piezo drive (Reproduced by permission of Apex Microtechnologies Inc.,USA) (a) Implementation (b) Photograph

278 POWER ELECTRONICS DESIGN HANDBOOK

TABLE 10-1 Some Representative Power Hybrid Circuits

Company	Part No.	Circuit Type	Max Ratings	Features
Apex µ-Tech	PA03	Power Op Amp	75V, 30 A	1mhz Gain-Bandwidth Product, 30 KHZ Power Bandwidth, 8v/µsec Slew Rate.
Gentron	SSR Series	Solid-State Relays	1200V, 25-125A	Uses MOSFET or IGBT Switches
	GS-105	Half-Bridge Motor Driver	400V, 25A	Simultaneous-Conduction Lockout, Over Current Protection, Isolated Package.
ILC Data Device	PWR-82331	3-Phase Bridge Motor Driver	200V, 30A	Mil-Std-883C Process, Six MOSFETS, Six Diodes, Digital Control, and Protection Circuitry
	SSP-21110	Solid-State Power Controller	28V, 2-25A	Contains High-Side MOSFET Switch, Driver, and Isolated Controller
Modupower	Mp7000 Series	Buck Regulators	15V, 1A	Combines 5-Terminal Regulator with Magnetic and Filtering Elements
Motorola	MPM3002	MOSFET Half-Bridge	100V, 8A	Icepak Power Modules for Single and 3-Phase Motor Drive. Used With Company's MCxxxx Series Of Control ICs
	MPM3003	MOSFET 3-Phase Bridge Circuit	60V, 10A	
Omnirel	OM9011SF	Power Module	100V, 18A	Mtl-Std-883C Process, Four MOSFETS, Four Rectifiers, and Four Schottky Diodes
SGS-Thomson	GS-R400V	Step-Down Switching Regulator	5 to 40V, 4A	Programmable Output Voltage, Soft-Start, Short-Circuit, and Thermal Protection
SIPEX	SP2805	DC/DC Converter	5V, 10A	1MHZ Operation, Has Input/Output Filters, Transformer, Switching Transistor, and Control

10.6 Smart Power Devices

The term "smart power" is difficult to define. Because of its catchy name, vendors have applied the term to products that only marginally qualify as smart-power devices. Some definitions specify an output voltage rating of 50V or greater without regard to the device's current rating. Any reasonable definition of smart-power specifies a power-handling capability of at least 1W. Far more difficult to define than a minimum power rating is the amount of intelligence or "smarts" a device needs to qualify as being smart. The intelligence can be in the form of control functions, interface capability, or fault management.

The basic idea underlying smart power technologies is to modify MOSFET or bipolar transistor technology to create islands for digital and analog signal processing functions. If possible, the signal section ICs should not reduce the performance of the power section.

Despite the lack of a commonly accepted definition, smart-power ICs represent a technology that is applicable to many of today's system requirements. These intelligent circuits ease the interfacing of digital and analog ICs with discrete power devices. With most of the required circuitry contained in a single package, both monolithic and hybrid smart-power ICs offer system designers the benefits of reduced size, greater reliability, and protection. This last benefit is often where a smart-power device really demonstrates superior intelligence by protecting itself against potential catastrophes such as voltage over-loads, short circuits, and thermal runaway.

When designing a smart-power integrated circuit, several issues that are not of great importance in standard low-power designs need be addressed. Among the most important issues are current, voltage, and safe operating area of the power transistors in the IC. Other considerations are proper layout, length, and width of the high-current interconnect lines. Also important is the proper design of the intermediate or drive stages for the power transistors and the effects of heat generated in the power transistors on the small signal circuitry in addition to proper design and layout of the circuitry. The proper choice of parallel and series connections of resistors with respect to their ohmic value and power requirements is also important. Finally, there is the proper choice of an IC package with respect to the desired number of pins and required power dissipation, as well as the proper IC layout to obtain a bondable device.

In contrast to a PIC, a smart device is a discrete component with additional features integrated into it. Various specialities may be built-in, like sensing of voltage, current, temperature, di/dt, dV/dt, etc. Usually, the goal is the protection of either the device or the application from unwanted events. The future potential of these devices is closely connected to the cost of an application specific design. Typically, the switching power is somewhat higher than that of a PIC of equal chip size.

One category of smart power technology is classified as "vertical" technologies (Figure 10–4(b)). The load current enters the die through the backside metallization, is controlled by vertical transistor structures and leaves the die trough the upper side metallization exactly as in discrete transistors. Current/voltage ratings and performances of power stages are close or equal to optimized discrete power semiconductor devices. Multiple high-side switches (with common drain) can be made with these technologies but multiple low-side switches are not feasible.

With "lateral" smart power technologies the load current enters the die through the upper side metallization (Figure 10–4 (a)). Current is controlled by lateral transistor structures and leaves the die again through the upper side metallization. The on-resistance per area unit is dependent on blocking voltage, with high voltage devices requiring excessive die size. Lateral technologies are suitable for integration of low voltage multiswitch circuits.

In "quasi-vertical" technologies the current flow is controlled by power stages with vertical current flow in the drain or collector layer (Figure 10–4(c)). It enters the die through the upper side metallization and is conducted downwards by means of a "sinker." The sinker connects to a buried layer with lateral current flow, distributing current to the MOSFET's drain. Sinker and buried layer increase on-resistance and the upper side drain contact requires extra die area, therefore limiting performance. Quasi-vertical technologies are suitable for applications where isolation between substrate and active area is required and for multiple low-side switches, high-side switches, half-and full-bridges.

For details Rischmuller (August and September 1992) are suggested.

10.6.1 Smart Power Hybrids

The recent interest in power hybrids has its beginnings in packaging multi-chip power devices. Actually, the multi-chip modules (MCM) are not considered as hybrids, and require external components to complete a function. The MCMs afford higher density and offer a level of design simplification. The MCMs and simple power hybrids herald the emergence of a new product category: "smart-power hybrids." Presently, custom designs seem to dominate those smart-power hybrids in use, and virtually all invoice formidable, very specific criteria regarding mechanical (especially volume and weight), thermal, and circuit performance limitations.

10.6.2 Custom Smart Power ICs

In addition to the multitude of standard smart power devices, you have the option of purchasing a custom device—provided the expected production quantities justify the development costs. Indeed, many system designers with large volume applications take the custom route rather than use a standard device, which may not

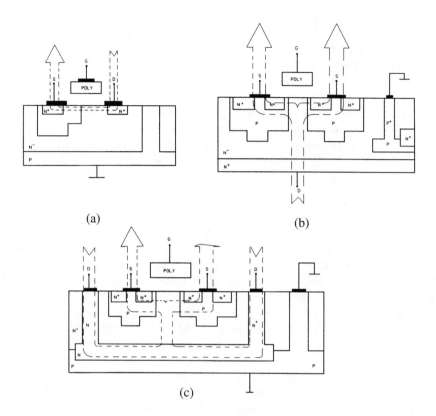

FIGURE 10-4 Simplified classification of basic smart power technologies (a) Lateral (b) Vertical (c) Quasi vertical (Reprinted with permission from *PCIM Magazine*)

provide the exact performance they need. In many cases, however, a semi-custom circuit can fit the bill.

10.6.3 Semi-Custom Smart Power ICs

When quantities cannot justify the cost of a custom circuit and a standard product doesn't quite meet the designer's objectives, a semi-custom circuit can often bridge the cost-performance gap. The design of a semi-custom smart power IC is normally done using the following steps:

(a) System design in the form of a block diagram.
(b) Circuit design for each block in the diagram.
(c) Computer simulation of the designed circuitry, using a simulation program such as SPICE, with the proper model parameters of the various components in the circuit, in particular the power devices.
(d) Modification of the design and additional simulation to verify proper performance.

(e) Additional circuit simulation at the entire temperature range of the device, especially at high temperatures which may be a result of the power dissipated in the power transistors.
(f) Construction of a breadboard and the use of proper "kit parts."
(g) Testing the breadboard at the entire temperature range, especially at elevated temperatures.
(h) Layout of the semi-custom IC, on the properly chosen smart power array.
(i) Generation of two test programs, one for testing the silicon wafers at much reduced current levels and another test program for testing the packaged devices at the desired power levels, in accordance with the specific application.

The 6000 series of semi-custom devices from Cherry Semiconductor Corp. was a good example of the inherent versatility of this type of device. The 6000 array contains two monolithic chips in either a 15-lead multi-watt package or 20- or 24-lead small outline (SO) package (Figure 10–5). One of the chips integrates two power npn transistors, each rated at 50V and 2.5 A. The transistors feature a typical h_{FE} of 80 and a V_{CE} (sat) of 0.75V at a collector current of 2A. Connected in parallel, the two transistors can sink as much as 5A to an external load.

Solitron's semi-custom devices include several array sizes for operating voltages of 20V, 40V, and 80V. Obviously, the smaller arrays include a lower number of small signal components as well as power devices with a lower current rating. Each array is about 50 percent larger than the previous size array and has output devices rated for about twice the current rating of the previous smaller array. These arrays contain: low power NPN, Schottky PNP, high power NPN, low power PNP and medium power PNP transistors, zener diodes, Schottky diodes, capacitors, and resistors.

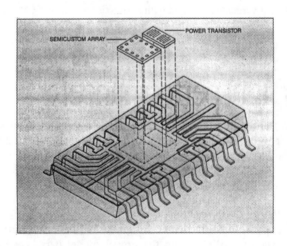

FIGURE 10–5 Genesis 6000 semi-custom array (Reproduced with permission by Cherry Semiconductor Inc., USA)

The semi-custom array of Figure 10–6 is typical of the rest of the family. Along one edge of the IC, four power NPN transistors occupy about one-third of the die area: small signal components fill the rest. This layout provides good separation between heat-generating and small signal processing components. A designer can use this layout to minimize the thermal effects on small signal components by using balancing techniques, such as cross coupling or centering components.

Along the edges of the chip are 10 x 10 mil bonding pads used for heavy aluminum wires that carry high current from the package pins to the power transistors. For practical rather than technical reasons, the IC's small signal portion interfaces with similar bonding pads, allowing a single bonding machine to service the entire IC. The silicon area under each bonding pad contains a 7V zener diode that can also function as a 15pF junction capacitor. The area next to the power NPN transistors in the array contains some medium size NPN and PNP transistors. You can use these to form Darlington or even double Darlington connections as well as to form composite PNP connections.

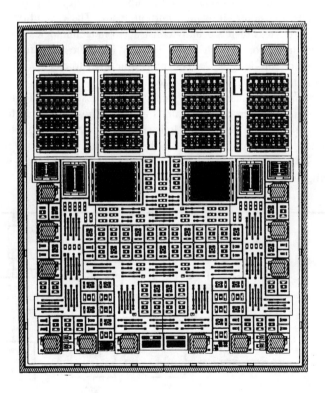

FIGURE 10–6 A photograph of 40V smart power semi-custom array (Reprinted with permission from *PCIM Magazine*)

284 POWER ELECTRONICS DESIGN HANDBOOK

The latter is made with a large size PNP transistor that drives a power NPN transistor. The small signal transistors, located in the lower section of the IC, are normally used to form the control circuitry of the smart power design. This circuitry may include digital functions such as logic gates and flip-flops, or analog functions such as operational amplifiers and voltage regulators. An adequate number of resistors, with a binary scale of nominal values, is provided in the vicinity of each transistor to minimize the effort of interconnecting and routing. A predetermined grid is positioned over the entire array area for the purposes of tracing the routing lines.

Using a semi-custom device such as W40E from Solitron, one can implement a smart power relay driver that includes the digital and analog control circuitry as well as the power drive stage, on the same semi-custom IC chip. Figure 10–7 shows the block diagram of a smart power relay driver.

10.7 Smart Power Microcontrollers

In advanced motion control systems, etc. a single-chip microcontroller or microprocessor chip set is usually teamed up with devices capable of processing analog signals and controlling high current loads. These additional devices are either discrete devices or ICs usually constructed in technologies dissimilar to that of the controller IC or to each other. To fully integrate these functions, new IC processes and circuit designs were developed in order to create a single-chip solution for motion and power control applications.

The merging of analog power and microprocessor technologies on a single chip has resulted in microcontroller units (MCU). Those enhanced towards power control are called Power Controller Units (PCU). By 1990 an initial test device had been produced by Motorola to evaluate this integrated technology (Artusi, Jorvig, Shaw, and Sutor 1990). At least four such controllers from Motorola were commercially available by the end of year 1996.

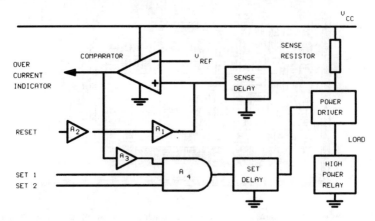

FIGURE 10–7 Smart power relay driver

10.7.1 Architecture of a PCU

The system level block diagram of the PCU is shown in Figure 10–8 (Artusi, Jorvig, Shaw, and Sutor 1990). The main controlling function is provided by the popular 8-bit 68HC05 CPU core supported with 96 bytes of RAM and 2064 bytes of user ROM. An additional 240 bytes of ROM are present to support production testing and self-test functions.

This CPU is a well-known industry standard that contains an 8-bit accumulator and an 8-bit index register. It supports 65 instructions and 10 addressing modes, including an 8x8 hardware multiply, bit testing and bit manipulation. This CPU can support seven levels of interrupts and 64K of memory-addressing space. One CPU bus cycle lasts two cycles of the on-chip oscillator with a maximum bus speed of 4MHz, which yields a typical instruction execution time of less than 1 µsec.

The two programmable 8-bit I/O ports provide a total of 16 pins, which can be configured as either inputs or outputs using a software programmable data direction register. The multipurpose timer provides several timing functions that include an 8-bit free-running timer with overflow interrupt, a prescalable real-time interrupt and a computer operating properly (COP) detector.

The timer is clocked by 1/4 th frequency of the CPU bus clock for a maximum count speed of 1 MHz. The real-time interrupt and COP function are driven by the

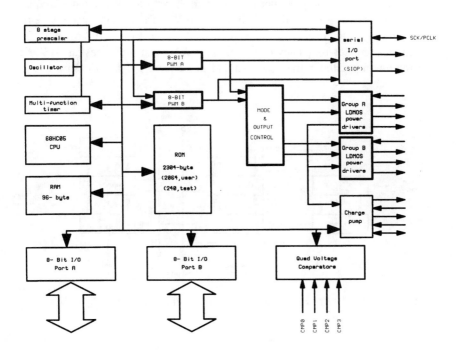

FIGURE 10–8 PCU Overall block diagram (Copyright of Motorola, used by permission)

8-bit free-running timer overflow through a 4-stage software programmable prescaler, which provides the following timing for a 4MHz CPU bus clock:

- The free-running timer overflows every 256 μsec.
- The real-time interrupt is generated every 4, 8, 16 or 32 milliseconds.
- The COP will reset the CPU in 33, 66, 131, or 262 millisec, if not properly serviced by the software.

The serial I/O port (SIOP) is a simple serial interface with a synchronous 8-bit format and simultaneous transmit and receive registers connected to separate pins for data output and data input. The data clock is bidirectional depending on whether the SIOP is programmed as a master or slave device. The order of transmission (MSB or LSB first) and the clock baud rate can be specified with the software code for the ROM.

Two pulse-width modulators (PWM) with a common prescaled clock source are provided, as detailed in Figure 10–9. These PWMs are implemented in hardware to eliminate the constant attention required by most software-driven PWM schemes. Each PWM can be programmed with an 8-bit number to provide duty-cycles in steps of 0.4 percent rom 0 to 99.6 percent (255/256).

The frequency of operation of both PWMs is set by a single programmable 8-stage prescaler that is clocked from the on-chip oscillator. The maximum PWM frequency occurs when the prescaler is connected directly to the oscillator frequency (divide by 256 overall) and the minimum PWM frequency occurs when the prescaler divides the oscillator frequency by 128 (divide by 32768 overall). For a maximum oscillator frequency of 8MHz the PWMs can run as fast as 31.25KHz. The minimum oscillator frequency should be at least 1MHz to ensure adequate performance.

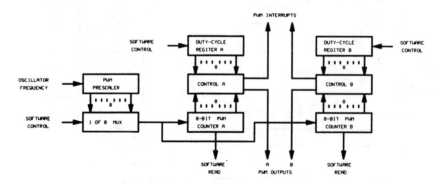

FIGURE 10–9   PWM Block diagram (Copyright of Motorola, used by permission)

10.7.2 Commercially Available Devices

Commercially available devices based on the above technology by Motorola are 68HC705MC4, M68HC708MP16, MC68HC16Y1, and MC68332G. The first two of these silicon efficient, cost effective controllers are for brushless DC motors and three phase induction motors, respectively. If a highly flexible general purpose machine is preferred MC68HC16Y1 and MC68332G could be used. These two flexible Time Processing Units (TPU) with 16 channels are actually having separate processor cores for complex timing tasks. Table 10–2 compares the four devices.

TABLE 10–2 Comparison of Motorola Motor Control Units (Source: Motorola)

	68HC705MC4	68HC708MP16	MC68HC16Y1	MC68332G
Processor core	HC05	HC08	HC 11	68000
ROM/EPROM	3.5K	16K	48K	None
RAM	176	512	2K	2K
Timer	16-Bit	16-Bit	16 Ch TPU	16 Ch TPU
A to D converters	6 channel (8-bit)	10 channel (8-bit)	10 channel (8-bit)	None
PWM	2 channel (8-bit, 23.4Khz Max)	6 channel (12-bit)	16 Channel TPU	16 Channel TPU
I/O	22	37	95	47
Self Check	COP	COP	WDOG	WDOG
Application	Brushless DC motor	3φ Induction motors	General purpose motor control	General purpose motor control

10.7.2.1 68HC705MC4

The 68HC705MC4 is an HC05-based MCU designed for three-phase brushless DC motor (permanent magnet) drive applications. General features include 3.5 Kbytes of EPROM, 176 bytes of RAM, a 16-bit timer, 4 general-purpose I/O pins, and an SCI (UART) port in a 28-pin SOIC or DIP package. In addition the MC4 has specific features that target brushless DC motor control including a 2-channel, 8-bit PWM module; a high current source port; and a 6-channel, 8-bit A/D module. Key features of the 6-pin, 2-channel PWM module include: 16 PWM rates between 122Hz and 23.4KHz for each PWM channel, buffered data registers with an interlocking mechanism for coherent updates of the pulse width outputs on each PWM channel, and a 3 output commutation MUX connected to an output port with 10 mA current sink capability per pin (thus cost reducing the external components required for building motor drives).

10.7.2.2 68HC708MP16

The 68HC708MP 16 is an HCO8-based MCU designed for open loop three-phase AC induction motor drive applications. General features include 16 Kbytes of ROM, 512 bytes of RAM, 2 16-bit timers, SPI, SCI, (UART), 13 general-purpose I/O, and an LVR module in a 64-pin QFP package. The MP16 also has specific features that target AC induction motor applications including a 6-channel, 12-bit PWM module; a high current sink port; and a 10-channel, 8-bit A/D module. Key features of the 6-channel PWM module include center- or edge-aligned modes, a mode that configures the six outputs as complementary pairs for coherent updates, a dead-time generation register to prevent shoot through currents in the motor drive circuit, current sense pins to correct for dead time distortion, and fault detect pins for fast shut down of the PWM outputs.

The hardware contained in the PWM module eliminates the need for several external components (i.e. logic for current sense, deadtime generation, and fault handling). Overall the MP16 offers high performance for an affordable price in open loop AC induction motor applications.

10.7.2.3 MC 68HC16Y1 and MC 68332G

Timing intensive signals common to motion control applications are easily handled by Time Processing Units (TPUs) in the MC68HC16Y1 and MC68332G. These 16 channel TPUs are actually separate processor cores dedicated to performing complex timing tasks. They handle a wide variety of motor and motion control tasks with great flexibility. TPU functions such as Hall Effect Decode, Quadrature Decode, Multiphase Commutation, and Stepping facilitate the implementation of motor control designs. Bursky (1996) provides details about 8-bit MCU for automotive system with high current and high voltage drive capability.

10.8 System Components and Impact of IGBTs

For building intelligent modules, in principle any turn-off power semiconductor can be used. Therefore, quality control of specific characteristics and a rough estimation of costs for triggering and protection are required first. One can see that the MOSFET and IGBT, which can be triggered almost without any power, offer the very best conditions for an advanced module construction. Their switching characteristics also are more favorable, so that they secure low-loss applications for quite a wide frequency range.

The reduced safe operating area of bipolar elements appears to be a disadvantage when discussing hybrid protection concepts. MOS-devices, however, can be protected by simple and nearly powerless circuits. In addition, they offer the pos-

Power Integrated Circuits, Power Hybrids, and Intelligent Power Modules **289**

sibility of chip-integrated sensors. The IGBT, which combines above mentioned advantages with low on-state values of bipolar switches for a wide power range, currently is popular for the construction of intelligent power modules. Practical examples of Intelligent Power Modules (IPM) using IGBTs are the Intellimod-3 from Powerex, Inc. USA (Motto and Williams, August and September 1992), ISO-MART from IXYS Semiconductors (IXYS Semiconductors, Technical Information 36).

Figure 10–10 shows the IPM's internally integrated functions and the isolated interface circuits and control power supply that the user must provide. The internal gate control circuit requires only a simple +15V DC supply. Specially designed gate

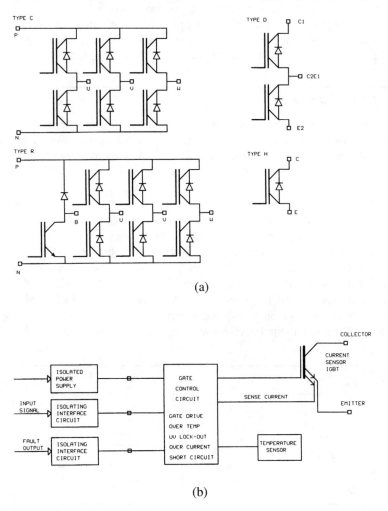

FIGURE 10–10 Intellimod family (Courtesy of Powerex Inc, USA) (a) Configurations (b) Block diagram

drive circuits eliminate the need for a negative supply to off bias the IGBT. The IPM's control input can interface with opto-coupled transistors with a minimum of external components.

Intellimod devices have ratings from 10A to 600 A at 600V and 10 to 300A at 1200V. Figure 10–10(a) shows available configurations: C, D, R, and H.

A block diagram of an ISOMART half-bridge module is shown in Figure 10–11. The IGBTs are driven by a digital 15V CMOS interface, which is galvanically isolated from the main terminals by toroid pulse transformers. On the right side, one finds the IGBTs with their corresponding free wheeling diode in phase-leg configuration. A specific driver IC (ASIC) includes a low loss gate drive circuit and the control of the gate-and collector voltage of the IGBT.

The energy needs by the ASICs is supplied by a controlled SMPS on the logic level side of the IPM. The energy is provided via a toroid transformer. For details on IXYS Semiconductors, Technical Information 36 is suggested.

10.9 Future

For now, most of the action seems to be in medium-level integration, with an emphasis on keeping costs down and efficiently implementing functions, either using a discrete transistor with protection or using a more complex control IC. However, commercialization of these products take 5–10 years of research and development.

Meanwhile, some companies are working on technology to combine BCD processes to better compete with discrete approaches. For example, Siliconix Inc., USA recently unveiled a line of power MOS-FETs that uses a patented vertical-trench structure. If the company could combine such vertical structures with CMOS and bipolar structures (it already has a BCD process), many of the cost and performance tradeoffs between discrete and integrated transistors would disappear. Another interesting development is BCD3 technology, which can integrate EEPROMs. In the future, processes that combine programmability with power may provide the necessary design flexibility to tip the scale toward more complex power ICs.

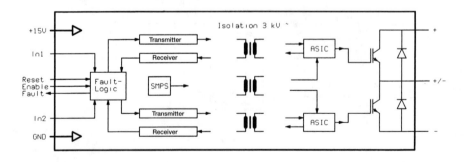

FIGURE 10–11 Block diagram of an ISOMART module (Courtesy of IXYS Semiconductors)

References

1. Pryce, D. "Smart Power ICs—Off the Shelf Circuits Increase in Popularity." *EDN*, 21 June 1990, pp 113–122.
2. Friedman, N. "Semi-custom, Smart Power IC Family Operators at 20V, 40V, 80V." *PCIM*, March 1988, pp 15–20.
3. Hopkins, T. "Smart Power—Technologies for All Applications." *PCIM*, February 1990, pp 19–24.
4. Antognett, P. *Power Integrated Circuits*. McGraw-Hill, 1986.
5. Danz, G. and T. Ferguson. "Power ASIC Simplifies UPS Design." *PCIM*, August 1990, pp 25–29.
6. Davis, S. "Cell-Based ASICs Allow User-Designed Intelligent Power Devices." *PCIM*, July 1990, pp 10–16.
7. Artusi, D., J. R. Jorvig, and M. L. Shaw. "Industries First Monolithic Smart Power Microcontroller Handles Over Six Watts." *PCIM*, November 1990, pp 12–18.
8. Artusi, D., J. R. Jorvig, M. L. Shaw, and J. Sutor. "The Ultimate Power IC for Intelligent Motion Control." Power Conversion International Proceedings, October 1990, pp 131–143.
9. Wodarczyk, P. J. and J. E. Wojslawowicz. "Intelligent Discretes: A New Era in Power Devices." Power Conversion International Proceedings, October 1990, pp 100–111.
10. Emerald, P. R., M. T. Hickey, and A. D. Tasker. "Perspectives on SMART Power Hybrids." Power Conversion International Proceedings, September 1991.
11. Jaecklin, A. A. "Future Devices and Modules for Power Electronic Applications." EPE–93 Proceedings, September 1993, pp 1–8.
12. Pryce, D. "Multi-Chip Circuits Satisfy Special Needs." *EDN*, 21 January 1991, pp 77–82.
13. Schulze, G. and M. Tscharm. "The Technique of Intelligent Modules." *EPE Journal*, June 1994, pp 27–32.
14. Schofield, G. "The Advantages of IC Power OP Amps." High Performance Amplifier Handbook (Vol. 1V), Apex U Tech Inc., pp 11–12.
15. Swager, A. W. "Power ICs: Weighing the Benefits of Integration." *EDN*, 7 July 1994, pp 68–82.
16. Kerridge, B. "Intelligent Power IC: Applications Drive Up Single Chips' IQ." *EDN*, 17 March 1994, pp 27–33.
17. Sax, H. "Intelligent Power MOSFET Protects Itself." *PCIM*, March 1995, pp 50–55.
18. Rischmuller, K. G. "Monolithic Smart Power Technology Moves Toward Higher Performance—Part I." *PCIM*, August 1992, pp 6–10.
19. Rischmuller, K. G. "Monolithic Smart Power Technology Moves Toward Higher Performance—Part II: Applications." *PCIM*, September 1992, pp 53–55.
20. Motto, E. and R. Williams. "IGBT-Based Intelligent Power Modules Reach 1200V/300A—Part I." *PCIM*, August 1992, pp 46–49.
21. Motto, E. and R. Williams. "IGBT-Based Intelligent Power Modules Reach 1200V/300A—Part II: Interfacing." *PCIM*, September 1992, pp 20–26.
22. Zacher, B., J. Pyle, and W. Weiss. "Monolithic 3.4A, 45V Intelligent Power IC Operates as Print Hammer or Motor Driver." *PCIM*, October 1992, pp 40–47.
23. Gauen, K. "Mixed Technology Power ICs Improve Circuit Performance and Reliability." *PCIM*, August 1993, pp 10–18.
24. SGS Thomson Microelectronics. "Advanced "BCD" Technology Brings Microcontroller Cores and Non-volatile Memory to Smart Power Chips." *EPE Journal*, March 1994, pp 5–6.
25. Devore, J., R. Teggatz, and C. Compton. "Monolithic Power IC Uses Nonvolatile Memory to Increase System Flexibility." *PCIM*, May 1994, pp 49–56.
26. Kanner, J. and K. Wellnitz. "Power IC Controls Automobile's Electronic Fuel Injectors." *PCIM*, February 1995, pp 24–31.
27. Powers, C. "On-Chip Voltage Regulator Cuts Microcontroller System Parts Count." *PCIM*, March 1995, pp 68–72.
28. IXYS Semiconductors. "Isosmart IGBT Modules Intelligent Power Modules (IPM) with Integrated Galvanic Isolation Interface." Technical Information 36 (D 94005 E).
29. Bursky, D. "Integrated 8-Bit MCU Handles High-Power Applications." *Electronics Design*, December 2, 1996, pp 85–89.

Index

AC-to-DC converters, 1
AGM (Absorbed Glass Mat), 143
Alkaline batteries, reusable, 156–57
   cumulative capacity, 157
Amplifiers, magnetic, 119–21
API (Application Programming Interface), 172
Applications, specialized, 6–8
Architecture, master/slave, 89
Architecture of PCU, 285–86
ASCR (Asymmetrical Thyristors), 25
ASDs (Adjustable Speed Drives), 247
ASICs (Application Specific Integrated Circuits), 230
ASPIC (Application Specific Power Integrated Circuits), 4

Ballast factor defined, 234
Ballasts, 229–42
   conventional, 235–36
   dimming, 239
   electronic, 229–42
   future developments of electronic, 241–42
   high frequency resonant, 236–37
   introduction to, 232–33
   magnetic and electronic, 239–41
   next generation of, 237–39
   and power factor correction, 239
Bandwidth control loop, 87–88
Batteries, rechargeable, 137–74
   introduction, 137
   lead-acid batteries, 142–48
      charging, 146–48
      flooded-lead-acid batteries, 142–43
      overcharging, 148
      sealed-lead-acid batteries, 143–46
   Li-Ion (lithium-ion) batteries, 153–56
      charge characteristics, 155–56
      construction, 154–55
      discharge characteristics, 155–56
   management, 159–73
      battery health, 169
      charging systems, 160–68
      End of Discharge (EOD) determination, 168
      gas gauging, 169
      semiconductor components, 170–72
      Smart Battery Data (SBD) specifications, 172
      Systems Management Bus (SMBus), 172
   NiCd (nickel cadmium) batteries, 148–51
      charge characteristics, 149–51
      construction, 148
      discharge characteristics, 148
      voltage depression effect, 151
   nickel metal hydride batteries, 151–53
      comparison between NiCd and NiMH batteries, 152–53
      construction, 152
   reusable alkaline batteries, 156–57
      cumulative capacity, 157
   technologies, 141–42
   terminology, 137–41
      C rate, 139
      capacity, 137–38
      charge acceptance, 140
      cycle life, 139–41
      cyclic energy density, 140
      depth of discharge, 140
      energy density, 139
      midpoint voltage, 140
      overcharge, 141
      self-discharge rate, 140
      voltage plateau, 140
   and their management, 137–74
   Zn-air batteries, 158–59
Battery
   chargers, 212–13
      converters for, 85–86
   health, 169
Battery management, 159–73
   charging systems, 160–68
      charge termination methods, 160–62
      Li-Ion chargers, 167–68
      NiCd and NiMH fast charge methods, 163–65
      sealed lead-acid batteries, 165–66
      temperature termination methods, 160–61
      voltage termination methods, 161–62
   EOD (End of Discharge) determination, 168
   SBD (Smart Battery Data) specifications, 172
   semiconductor components, 170–72
   SMBus (Systems Management Bus), 172
Battery terminology
   capacity, 137–38
      actual capacity, 138
      available capacity, 138

294 POWER ELECTRONICS DESIGN HANDBOOK

rated capacity, 138
retained capacity, 138
standard capacity, 138
Battery-operated equipment, converters for, 85–86
BCD (Bipolar-CMOS-DMOS) technology, 274–75
BEF (ballast efficacy factor) defined, 234
Bipolar power transistors, 28–36
BJT (bipolar junction power transistors), 9, 31, 33, 35–36
Board level protection, 182
Boost
  converters, 61–62
  inductors, 255
  topologies, 252–55
Buck-boost
  converters, 62
  topologies, 255–56

C rate, 139
Capacitance, junction, 17–18
Capacitor converters, switched, 90–92
Capacitors, 102, 127–29
  output, 255
Capacity, 137–38
CDM (Charged Device Model), 186
CFL (compact fluorescent lamps), 231, 240
Charge acceptance, 140
Charge termination methods, 160–62
Chargers, battery, 212–13
Charging systems, 160–68
  charge termination methods, 160–62
  NiCd and NiMH fast charge methods, 163–65
  sealed lead-acid batteries, 165–66
  temperature termination methods, 160–61
  voltage termination methods, 161–62
Circuits
  ancillary, supervisory, and peripheral, 129
  protection of, 186–88
CMOS (complementary metal-oxide semiconductor), 2–3, 36
Components
  modern, 4–6
  semiconductor, 3–4, 170–72
Computers, power supplies for, 129–30
Conditioners, line, 199
Converters
  DC to DC, 55–98
  for battery chargers, 85–86
  for battery-operated equipment, 85–86
  multi-stage off-line, 256–57
  topologies, 60–70
    flyback converters, 64–65
    full-bridge converter, 66–67

half-bridge converter, 66
non-transformer-isolated topologies, 60–63
push-pull converters, 65–66
summary and comparison, 69–70
transformer-isolated Cuk converters, 67–68
transformer-isolated topologies, 63–68
COP (Computer Operating Properly) detectors, 285
Core geometry, 109–12
Core materials, common, 109–12
CP (Constant Potential), 165
CSA (Canadian Standards Association), 100
Cuk converters, 63
  transformer-isolated, 67–68
Current
  crest factor defined, 234
  inrush, 103
  leakage, 17–18
Current-mode controller, typical, 74–76
CVTs (Constant Voltage Transformers), 189
Cycle life, 139–41
Cyclic energy density, 140

Darlington transistors, 35–36
DC to DC converters, 55–98
  analysis, 59–60
    volt-second balance for inductors, 59–60
  applications, 93–94
  and ICs, 93–95
  control integrated circuits, 74–76
    typical current-mode controllers, 74–76
    typical voltage-mode controllers, 74
  control of, 71–76
    PWM control techniques, 71–73
  converter topologies, 60–70
    flyback converters, 64–65
    full-bridge converter, 66–67
    half-bridge converter, 66
    non-transformer-isolated topologies, 60–63
    push-pull converters, 65–66
    summary and comparison, 69–70
    transformer-isolated Cuk converters, 67–68
    transformer-isolated topologies, 63–68
  design, 84–93
    converters for battery chargers, 85–86
    converters for battery-operated equipment, 85–86
    practical design approaches, 86–93
    sub-5V applications, 84–85
  fundamentals, 56–60
  future directions, 95–96
  ICs, 94–95
  introduction, 55–56

modes of operation, 56–58
    flyback mode converters, 57–58
    forward mode converters, 56–57
practical design approaches, 86–93
    bandwidth control loop, 87–88
    idle mode control scheme, 89
    improved process technologies, 90
    increased gain, 87–88
    master/slave architecture, 89
    Single Ended Primary Inductance
Converter (SEPIC), 92–93
    switched capacitor converters, 90–92
    synchronous rectification, 86–87
    updated voltage mode control, 89–90
PWM control techniques, 71–73
    current-mode control, 72–73
    gated oscillator control, 73
    voltage-mode control, 71–72
resonant converters, 76–84
    control techniques, 82–84
    ICs for resonant converters, 82–84
    introduction, 76–77
    PWM vs. quasi-resonant techniques, 78
    quasi-resonant principle, 77
    quasi-resonant switching converters, 81–82
    resonant switches, 78–81
resonant switches
    ZC-QRCs (zero current quasi-resonant switches), 79–80
    ZV-QRCs (zero voltage quasi-resonant switches), 80–81
state of art, 95–96
DC-to-AC converters, 1
Density, cyclic energy, 140
Density, energy, 139
Detectors, COP (computer operating properly), 285
Devices, solid state, 180–81
Diodes, 102
    fast and very fast recovery, 121–22
    GaAs power, 4, 18–19
    power, 10–19
    Schottky, 17–18
Discharge
    depth of, 140
    electrostatic, 186–88
DMOS, 38
Drivers, gate, 222–23

EFT (electronic fast transients), 183
Electronic tap changers, 189
Electrostatic discharge, 182, 186–88
EMI/RFI (electromagnetic interference/radio frequency interference), 105, 132
Energy density, 139

cyclic, 140
EOD (end of discharge) determination, 168
EODV (end of discharge voltage), 138
ESD (electrostatic discharge), 182, 186–88
ESI (equivalent series inductance), 128
ESR (equivalent series resistance), 128

FCC (Federal Communications Commission), 105
Ferrites, 107–9
Ferroresonant regulators, 191–93
FETs (field-effect transistors), 36
Filters, RFI/EMI, 104–5
Fluorescent lamps, 230–31
Flyback
    converters, 64–65
    mode converters, 57–58
    transformers, 114–15
Forward
    mode converters, 56–57
    recovery, 10–12
FRED (Fast Recovery Epitaxial Diode), 122
Full-bridge converter, 66–67
Fuses, 102

GaAs power diodes, 4, 18–19
    semiconductor materials, 9
Gain, increased, 87–88
Gas
    discharge lamps, 230–32
    gauging, 169
    tubes, 180
Gate
    drive considerations, 40–42
    drivers, 222–23
GTOs (gate turn-off thyristors), 3–4, 9–10, 26–28

Half-bridge converter, 66
Harmonic control, 243–69
    harmonic standards, 248–49
        IEC 555-2, 248–49
    problems caused by harmonics, 247–48
Harmonics and power factor, 245–47
HBM (Human Body Model), 186
HID (high intensity discharge), 230–32

I-V characteristics, 39–40
ICs (integrated circuits)
    and DC to DC converters, 93–95
    ML series, 259–63
    custom smart power, 280–81
    MC33262, 263–65
    MC34262, 263–65
    new PWM, 116–17
    power, 271–91
    power applications specific, 4

power factor correction, 257–65
　for resonant converters, 82–84
　semi-custom smart power, 281–84
　UC3854, 257–59
ICs (integrated circuits), control, 74–76
　typical current-mode controllers, 74–76
　typical voltage-mode controllers, 74
Idle mode control scheme, 89
IEC (International Electrotechnical Commission), 100
IEEE C62.41-1991, 184–86
IGBTs (insulated gate bipolar transistors), 3, 7, 9, 47–49, 288–90
Inductive load switching, 30
Inductors, 106–15
　boost, 255
　and capacitors, 127–29
　volt-second balance for, 59–60
Inverter thyristors, 24
Inverters, 213–21
　practical circuits, 218–21
　selection of transistors, 217–18
　switching principles, 214–17
Isolation transformers, 195–98

JEDEC (Joint Electronic Device Engineering Council) titles, 25–26
JFETs (junction FETs), 36
Junction capacitance, 17–18

Lamps, 229–42
　compact fluorescent, 239–41
　definitions, 233–35
　　ballast factor, 234
　　BEF (ballast efficacy factor), 234
　　current crest factor, 234
　　luminous efficacy, 234
　　luminous flux, 233–34
　　THD (total harmonic distortion), 235
　energy saving, 229–42
　gas discharge, 230–32
　　CFL (compact fluorescent lamps), 231
　　fluorescent lamps, 230–31
　　HID (high intensity discharge), 230–32
　introduction, 229–30
Lead-acid batteries, 142–48
　charging, 146–48
　flooded-lead-acid batteries, 142–43
　overcharging, 148
　sealed, 143–46
　　capacity during battery life, 145
　　discharge performance, 143–44
　　effect of pulse discharge on capacity, 145–46
Leakage current, 17–18
Li-Ion (lithium-ion) batteries, 153–56
　charge characteristics, 155–56
　construction, 154–55
　discharge characteristics, 155–56
Light-triggered thyristors, 25
Line conditioners, 199
Luminous efficacy defined, 234
Luminous flux defined, 233–34

MAGLEV (magnetically levitated) trains, 7
Magnetic amplifiers, 119–21
Magnetic components
　new PWM ICs, 116–17
　planar magnetics, 116–17
Magnetic material, 106–9
　ferrites, 107–9
　magnetic metals, 106
Magnetic metals, 106
Magnetics, planar, 116–17
Master/slave architecture, 89
MCM (multi-chip modules), 280
MCT (MOS controlled thyristors), 4, 9, 49–52
Metals
　magnetic, 106
　powdered, 106–7
MG (motor generators), 199
Microcontrollers, use of, 221–23
Midpoint voltage, 140
MLC (multilayer ceramic), 128
MLP (multilayer polymer), 128
MM (Machine Model), 86
MOS controlled thyristors; See MCT (MOS controlled thyristors)
MOS (Metal-Oxide-Semiconductor), 3–4
MOSFETs (Metal-Oxide-Semiconductor Field Effect Transistors), 3–4
MOSFETs, power, 9, 36–47
　I-V characteristics, 39–40
　advanced, 46–47
　characteristics, 37–38
　gate drive considerations, 40–42
　introduction, 36–37
　practical components, 43–47
　　high and low voltage on resistance devices, 43–45
　　more advanced power MOSFETs, 46–47
　　P-channel MOSFETs, 45–46
　on resistance, 38
　safe operating area, 43
　structures, 38
　temperature characteristics, 42
Motor driven variacs, 188–89
MOVs (metal oxide varistors), 104, 180, 182, 188

NiCd (nickel cadmium)
　batteries, 148–51
　　charge characteristics, 149–51
　　construction, 148

discharge characteristics, 148
  voltage depression effect, 151
  fast charge methods, 163–65
Nickel metal hydride batteries, 151–53
  comparison between NiCd and NiMH batteries, 152–53
  construction, 152
NiMH fast charge methods, 163–65
Non-transformer-isolated topologies, 60–63

Output
  capacitors, 255
  rectification, 118–19
  voltage control, 221–23
Overcharge, 141

P-channel MOSFETs, 45–46
PASIC (Power Applications Specific Integrated Circuits), 4
PCU, architecture of, 285–86
PFC (power factor correction), 239, 243–69, 249–57
  active, 250–52
  active low frequency, 265–67
  definitions, 244–45
    power factor, 244
    THD (total harmonic distortion), 245
  evaluation circuits, 267
  harmonics and power factor, 245–47
  ICs, 257–65
    ML series, 259–63
    MC33262, 263–65
    MC34262, 263–65
    UC3854, 257–59
  introduction, 243–44
  multi-stage off-line converters, 256–57
  using boost topology, 252–55
    boost inductors, 255
    output capacitors, 255
  using buck-boost topology, 255–56
Phase controlled thyristors, 24
PIC (power integrated circuits), 271–91
  applications of, 275–76
  BCD (Bipolar-CMOS-DMOS) technology, 274–75
  evaluation of, 272–74
  future of, 290
  introduction, 271–72
PIV (peak inverse voltage), 10
Planar magnetics, 116–17
Powdered metals, 106–7
Power conversion electronics, 2
Power devices, smart, 279–84
Power diodes
  fast and ultra fast rectifiers, 15
  forward recovery, 10–12

GaAs power diodes, 18–19
  reverse recovery, 12–15
  Schottky rectifiers, 15–18
Power electronics
  importance in modern world, 2
  industry, 1
Power factor correction; *See* PFC (power factor correction)
Power factor defined, 244
Power hybrids, 271, 276–78
Power microelectronics, smart, 284–88
Power modules, intelligent, 271–91
Power MOSFETs, 9, 36–47
Power quality and modern components, 4–6
Power rectifiers, 121–27
Power semiconductors, 9–54
Power supplies; *See also* SMPS (switchmode power supplies)
  for computers, 129–30
  field trouble-shooting of, 133
Power synthesis equipment, 199
Primary protection, 181
Process technologies, improved, 90
Products, SNMP, 225
Protection
  board level, 182
  primary, 181
  secondary, 181–82
Protection systems, 175–200
  electrical noise, 176–77
    common-mode noise, 177
    normal-mode noise, 177
  introduction, 175
  for low power systems, 175–200
  for low voltage, 175–200
  power enhancement equipment, 178–88
    transient voltage surge suppressors (TVSS), 178–88
  power protection equipment, 177–99
    isolation transformers, 195–98
    line conditioners, 199
    power enhancement equipment, 178–88
    voltage regulators, 188–95
  power synthesis equipment, 199
  TVSS (transient voltage surge suppressors), 178–88
    performance considerations, 180–82
    practical considerations, 188
    practical surge protection circuits, 182
    practical TVSS, 180–82
  types of disturbances, 176–77
    blackouts, 177
    electrical noise, 176–77
    voltage sags, 176
    voltage surges, 176
    voltage transients, 176

voltage regulators, 188–95
   electronic tap changers, 189
   ferroresonant regulators, 191–93
   miscellaneous, 193–95
   motor driven variacs, 188–89
   thyristor driven AC regulators, 190–91
Push-pull converters, 65–66
PWM (Pulse Width Modulation), 213–14, 217–18, 221, 286
   control techniques, 71–73
      current-mode control, 72–73
      gated oscillator control, 73
      voltage-mode control, 71–72
   new ICs, 116–17
   vs. quasi-resonant techniques, 78

QSW (Quasi-Square-Wave), 213
Quasi-resonant principle, 77

RBSOA (reverse-bias safe operating area), 34–35
RCT (reverse conducting thyristor), 25
Rechargeable batteries, 137–74
Rectification
   output, 118–19
   synchronous, 86–87
Rectifiers
   and battery chargers, 212–13
   fast and ultra fast, 15
   power, 121–27
   Schottky, 15–18
   Schottky barrier, 122–23
   synchronous, 125–27
Regulators
   ferroresonant, 191–93
   thyristor driven AC, 190–91
   voltage, 188–95
Resistor-triac technique, 103
Resonant converters, 76–84
   control techniques, 82–84
   ICs for resonant converters, 82–84
   introduction, 76–77
   PWM vs. quasi-resonant techniques, 78
   quasi-resonant principle, 77
   quasi-resonant switching converters, 81–82
   resonant switches, 78–81
Resonant switches, 78–81
   ZC-QRCs (zero current quasi-resonant switches), 79–80
   ZV-QRCs (zero voltage quasi-resonant switches), 80–81
Reverse recovery, 12–15
Reverse-bias secondary breakdown, 34–35
RFI (radio frequency interference), 212
RFI/EMI (radio frequency interference/electromagnetic interference), 133
   filters, 104–5

SBD (Smart Battery Data) specifications, 172
Schottky
   barrier rectifiers, 122–23
   diodes
      junction capacitance, 17–18
      leakage current, 17–18
   rectifiers, 15–18
Secondary protection, 181–82
Self-discharge rate, 140
SELV (Safety Extra Low Voltage), 132
Semiconductor components, 3–4, 170–72
   ASPIC (Application Specific Power Integrated Circuits), 4
   CMOS, 3
   GaAs power diodes, 4
   GTO, 4
   GTOs (Gate Turn-Off Thyristors), 3
   IGBTs (Insulated Gate Bipolar Transistors), 3
   MCT (MOS Controlled Thyristors), 4
   MOS (Metal Oxide-Semiconductor), 3–4
   MOSFETs (Metal-Oxide-Semiconductor Field Effect Transistor), 3–4
   PASIC (Power Applications Specific Integrated Circuits), 4
Semiconductors, power, 9–54
   bipolar power transistors, 28–36
      V-I characteristics, 30–35
      Darlington transistors, 35–36
      inductive load switching, 30
      safe operating area, 30–35
      as switches, 28–30
   IGBTs (insulated gate bipolar transistors), 47–49
   introduction, 9–10
   MCT (MOS controlled thyristors), 49–52
   power diodes and thyristors, 10–26
      power diodes, 10–19
   power MOSFETs, 36–47
      I-V characteristics, 39–40
      characteristics, 37–38
      gate drive considerations, 40–42
      introduction, 36–37
      practical components, 43–47
      on resistance, 38
      safe operating area, 43
      structures, 38
      temperature characteristics, 42
SEPIC (Single Ended Primary Inductance Converter) converters, 92–93
Share mode defined, 205
SiC (Silicon Carbide), 9
SiO_2 (Silicon Dioxide), 37
SIOP (serial I/O port), 286
SLA (sealed lead-acid) batteries, 165–66
Smart power devices, 279–84
   custom smart power ICs, 280–81

Index 299

semi-custom smart power ICs, 281–84
smart power hybrids, 280
Smart power microelectronics, 284–88
   architecture of PCU, 285–86
   commercially available devices, 287–88
      68HC705MC4, 287–88
      68HC708MP16, 288
      MC 68HC16Y1, 288
      MC 68332G, 288
SMBus (Systems Management Bus), 172
SMES (superconducting magnetic energy storage), 5
SMPS (switchmode power supplies), 10, 99–135
   ancillary, supervisory, and peripheral circuits, 129
   field trouble-shooting of power supplies, 133
   future trends, 132; *See also* Power supplies
   high-frequency, 100–105
   inductors and capacitors, 127–29
      inductors and capacitors compared, 127–28
      output filter capacitors, 128–29
   input protective devices, 103–5
      input transient voltage protection, 104
      inrush current, 103
      resistor-triac technique, 103
      thermistor technique, 103–4
   input section, 100–103
      I^2t rating, 102–3
      capacitors, 102
      current ratings, 102
      diodes, 102
      fuses, 102
      let-through current, 102–3
      selection of basic components, 102
      voltage ratings, 102
   introduction, 99–100
   magnetic components, 106–17
      new PWM ICs, 116–17
      planar magnetics, 116–17
      transformers and inductors, 106–15
   modular SMPS units for industrial systems, 130–32
   off-the-line, 99–135
   output sections, 117–29
      filtering schemes, 118–19
      inductors and capacitors, 127–29
      magnetic amplifiers, 119–21
      output rectification, 118–19
      power rectifiers for switching power supplies, 121–27
      secondary side regulators, 119–21
   power rectifiers, 121–27
      fast and very fast recovery diodes, 121–22
      practical circuit considerations, 126–27
      Schottky barrier rectifiers, 122–23

   synchronous rectifiers, 125–27
   transient-over-voltage suppression, 123–25
   power supplies for computers, 129–30
   transformers and inductors, 106–15
      magnetic material, 106–9
SMT (surface mount technology), 128
SNMP products, 225
SOA (safe-operating area), 33
Solid state devices, 180–81
SPS (stand by power sources), 199, 202–3
Starved design defined, 143
Static switches, 223
Suppression, transient-over-voltage, 123–25
Surge protection circuits, practical, 182
Surge voltages in AC power circuits, 184–86
Switched capacitor converters, 90–92
Switches
   bipolar transistor as, 28–30
   resonant, 78–81
   static, 223
   ZC-QRCs (zero current quasi-resonant switches), 79–80
   ZV-QRCs (zero voltage quasi-resonant switches), 80–81
Switching
   inductive load, 30
   principles, 214–17
Switching converters, quasi-resonant, 81–82
Switchmode power supplies; *See* SMPS (switchmode power supplies)
Synchronous rectification, 86–87
Synchronous rectifiers, 125–27
Systems
   approaches, 6
   components, 288–90

Tap changers, electronic, 189
Temperature termination methods, 160–61
THD (total harmonic distortion), 235, 239, 245
Thermistor technique, 103–4
Thyristor driven AC regulators, 190–91
Thyristors, 10–26, 19–26
   ASCR (asymmetrical thyristors), 25
   different types of devices, 22–28
   GTOs (gate turn-off thyristors), 26–28
   inverter, 24
   JEDEC (Joint Electronic Device Engineering Council) titles, 25–26
   light-triggered, 25
   phase controlled, 24
   and popular names, 25–26
   ratings, 22–28
   RCT (reverse conducting thyristor), 25
TPUs (Time Processing Units), 287–88
Trains, MAGLEV (magnetically levitated), 7
Transformer-isolated Cuk converters, 67–68

Transformers and inductors, 106–15
  common core materials, 109–12
  core geometry, 109–12
  flyback transformers, 114–15
  inductors, 115
  magnetic material, 106–9
    ferrites, 107–9
    magnetic metals, 106
    powdered metals, 106–7
  practical considerations, 112–13
  transformers, 113–14
    isolation, 195–98
  winding techniques, 112–13
Transistors
  bipolar power, 28–36
    V-I characteristics, 30–35
    bipolar transistors as switches, 28–30
    Darlington transistors, 35–36
    inductive load switching, 30
    safe operating area, 30–35
  selection of, 217–18
Triport, 204–5
Tubes, gas, 180
TVSS (transient voltage surge suppressors), 178–88
  performance considerations, 180–82
  practical, 180–82
  practical considerations, 188
  practical surge protection circuits, 182

UL 1449, 186
UL (Underwriters Laboratories), 100
UPSs (uninterruptible power supplies), 199, 201–28, 247
  diagnostics, intelligence, and communications, 223–25
  future of, 226–27
  intelligence levels, 225
  intelligent systems, 224–25
  introduction, 201–2
  inverters, 213–21
    practical circuits, 218–21
    selection of transistors, 217–18
    switching principles, 214–17
  reliability, 226–27
  SNMP products, 225

  system components, 211–23
    battery chargers, 212–13
    gate drivers, 222–23
    inverters, 213–21
    output voltage control, 221–23
    rectifiers, 212–13
    static switches, 223
    use of microcontrollers, 221–23
  technology changes, 226–27
  types, 202–11
    hybrid UPS, 203–10
    line interactive UPS systems, 205–10
    off-line UPS, 202–3
    on-line UPS systems, 210–11

Variacs, motor driven, 188–89
VCO (voltage controlled oscillator), 82
VDE (Verband Deutscher Electronotechniker), 100, 105
VDMOS, 38
VIP (Vertical Integration Power) family, 132
Visually evaluated radiation, 233–34
Volt-second balance for inductors, 59–60
Voltage
  midpoint, 140
  plateau, 140
  protection, 104
  regulators, 188–95
    electronic tap changers, 189
    ferroresonant regulators, 191–93
    miscellaneous, 193–95
    motor driven variacs, 188–89
    thyristor driven AC regulators, 190–91
  termination methods, 161–62
Voltage control, output, 221–23
Voltage mode control, updated, 89–90
Voltage-mode controller, typical, 74
VRLA (Valve Regulated Lead Acid), 143

Winding techniques, 112–13

ZC-QRCs (zero current quasi-resonant switches), 79–80
Zn-air batteries, 158–59
ZV-QRCs (zero voltage quasi-resonant switches), 80–81